A TEXT BOOK

OF

CHEMISTRY.

A MODERN AND SYSTEMATIC EXPLANATION OF THE ELEMENTARY PRINCIPLES OF THE SCIENCE.

ADAPTED TO USE IN

HIGH SCHOOLS AND ACADEMIES.

BY

LEROY C. COOLEY, A.M.,

PROFESSOR OF NATURAL SCIENCE, IN THE NEW YORK STATE NORMAL SCHOOL,
AUTHOR OF A TEXT-BOOK OF NATURAL PHILOSOPHY.

NEW YORK:
CHARLES SCRIBNER & COMPANY.
1869.

PREFACE.

THIS volume is designed to be a *text-book* of Chemistry, suited to the wants of high schools and academies.

The author believes that the following features of his work adapt it to the purpose for which it was designed.

1. It contains no more than can be *mastered* by average classes in the time usually given to the study of Chemistry in the high schools and academies.

2. It is thoroughly systematized. The order and development of subjects is thought to be logical, and the arrangement of topics especially adapted to the best methods of conducting the exercises of the class-room.

3. It is written in accordance with modern theories, and no pains have been spared in the attempt to make it fairly represent the *present state* of the science as far as its elementary character will permit.

In addition to his attempt to make these features prominent, the author has not forgotten that a student will succeed best when required to learn *one thing at a time.* He believes that the difficulty often found by pupils in Chemistry does not lie in its laws, nor in its nomenclature, nor in its reactions, nor in any other one feature so much as in the illogical attempt to learn them all at once. He has therefore presented each one of these and other subjects separately and in natural order, like the successive steps of a ladder, leading to a height from which the pupil may have a clear view of the science.

Nor has he forgotten that Chemistry more than any other science rests upon experiment, that while its laws may be *explained* by certain theories, they are at the same time quite independent of such theories, being logical deductions from skillful and repeated experiment. He has sought to present them as such, and while the student is enlightened by the synopsis of the paragraph (numbered in parenthesis) concerning the object of the experiments which he is about to study, the law itself is made to appear as the result to which the experiments have led him. Moreover, while the properties of bodies may be illustrated by experiments made without especial precautions, laws can be established only by experiments from which all sources of error have been eliminated. To such the student's mind is directed.

The work is not designed to do away with oral instruction, but rather to facilitate it. The synopsis of the paragraphs are *texts* which, taken together, give an outline of the entire subject, and which the lecturer will find it profitable to illustrate by descriptions and experiments of his own in addition to those given in the topics which the student studies.

The author finds it impracticable to name all the authorities to which he is more or less indebted. He must, however, gratefully acknowledge the assistance derived from Hoffmann's Introduction to Modern Chemistry, Roscoe's Lessons in Elementary Chemistry, and Cooke's Chemical Philosophy, Part I.

For cuts Nos. 29, 30, 32, and 36, the author is indebted to Muspratt's Applied Chemistry; for Nos. 53 and 54, to Atkinson's Ganot's Physics (London, 1867); all others are from his own drawings.

L. C. C.

ALBANY, *June*, 1869.

ANALYTICAL CONTENTS.

INTRODUCTION.—ON PHYSICAL AND CHEMICAL CHANGES.

All changes to which matter is subject are either *physical* or *chemical*.

Chemistry is the science which treats of the composition of matter and the *chemical* changes to which it is subject.

CHAP. I.—ON THE COMPOSITION OF BODIES.

All substances are either elements, compounds, or mixtures.

ELEMENTS.—Nitrogen.
- Oxygen.
- Hydrogen.
- Carbon.
  - Nomenclature.
  - Symbols.

COMPOUNDS.—Pure water.
- Carbonic dioxide.
  - ANALYSIS.—By Electricity.
    - By the Prism.
    - By Chemical Action.
  - SYNTHESIS.

MIXTURES.—Air.
- Water.
  - Diffusion and Osmose.
  - Filtration, Evaporation, Distillation.

CHAP. II.—ON CHEMICAL ATTRACTION.

Attraction acting upon atoms of bodies under certain conditions and subject to definite laws produces all compound bodies.

- Conditions on which it acts.
- Laws which govern its action.
  - By Volume.
  - By Weight.
    - The Atomic Theory.
- The effect which it produces.
  - Compounds by Direct Combination.
  - Compounds by Substitution.
- Nomenclature of the compounds formed.
- Symbols representing composition and reactions.

CHAP. III.—ON CHEMICAL GROUPS.

Elements and compounds are so numerous that they must be studied in groups. That system is best which brings into the same group bodies whose properties are most nearly alike.

- The Non-Metals.
  - Quantivalence.
    - The Univalent or Chlorine group.
    - The Bivalent or Sulphur group.
    - The Trivalent or Nitrogen group.
    - The Quadrivalent or Carbon group.
- The Metals.
  - Metals of the Alkalies.
  - Metals of the Alkaline Earths.
  - Metals of the Earths.
  - The Zinc class.
  - The Iron class.
  - The Tin class.
  - The Tungsten class.
  - The Arsenic class.
  - The Lead class.
  - The Silver class.
  - The Gold class.

FRENCH MEASURES.

OF LENGTH.

1 Kilometre	=	1,000	Metre	=	0·6214	Mile.
1 Hectometre	=	100	"	=	328·09	Feet.
1 Decametre	=	10	"	=	32·809	"
1 METRE	=	1	"	=	3·2809	"
1 Decimetre	=	0·1	"	=	3·937	Inches
1 Centimetre	=	0·01	"	=	0·3937	"
1 Millimetre	=	0·001	"	=	0·03937	"

OF VOLUME.

1 Myrialitre	=	10,000	Litre	=	353·1659	Cub. Feet
1 Kilolitre	=	1,000	"	=	35·3165	" "
1 Hectolitre	=	100	"	=	3·5316	" "
1 Decalitre	=	10	"	=	0·3531	" "
1 LITRE	=	1	"	=	61·027	Cub. Inch.
1 Decilitre	=	0·1	"	=	6·1027	" "
1 Centilitre	=	0·01	"	=	0·61027	" "
1 Millilitre	=	0·001	"	=	0·061027	" "

OF WEIGHT.

1 Kilogramme	=	1,000	Gramme	=	2·20462	Pounds Avd.
1 Hectogramme	=	100	"	=	0·22046	" "
1 Decagramme	=	10	"	=	0·02204	" "
1 GRAMME	=	1	"	=	15·4323	Grains.
1 Decigramme	=	0·1	"	=	1·5432	"
1 Centigramme	=	0·01	"	=	0·1543	"
1 Milligramme	=	0·001	"	=	0·0154	"

---

1 Litre	=	1 Cubic Decimetre.
1 Decilitre	=	100 Cubic Centimetres.
1 Centilitre	=	10 " "
1 Millilitre	=	1 " "
1 Litre	=	0·22017 Gallon.
1 Crith	=	0·089578 Gramme.

CHEMISTRY.

INTRODUCTION.

PHYSICAL AND CHEMICAL CHANGES.

(1.) The changes which take place in bodies of matter are either physical or chemical.

1. *Physical changes.*—Bodies of matter are constantly changing. They move: how varied are their motions! They change their shapes, as when rocks are rounded by the flow of water over them, or shattered by a blast of gunpowder. They change from solid to liquid forms, and from liquids to gases, as when the snows of winter melt, or the dew of summer disappears. Such changes as these, however, do not affect the nature of bodies. After a body has moved, it is the same body as before. Water in the forms of ice and dew and vapor is water still. Changes like these, in which the nature of bodies is not affected, are called *physical changes.*

2. *Chemical changes.*—But all changes are not like these. Wood burns: in doing so it ceases to be wood; it is changed to smoke and ash. Gunpowder explodes: it is no longer gunpowder. Fluids from the soil and gases from the air are taken into the roots and leaves of plants, and are there changed into substances which

form the plant itself. Such as these are changes in the nature of substances. They are called *chemical changes.*

(2.) Chemistry is the science which treats of the properties and composition of matter, and of those phenomena in which there is a change in the nature of bodies.

1. *Chemistry and Natural Philosophy.*—Now, in the multitude of phenomena around us, we sometimes see changes taking place in the nature of bodies, and sometimes not. Those in which we do, are to be explained in Chemistry: those in which we do not, are subjects in Natural Philosophy. While the chemist has much to do with physical properties, he is chiefly interested with chemical changes, which he sees in nature, or which he brings about by his own operations, and by which he learns the composition of bodies and of various qualities of matter which physical changes cannot reveal.

2. *To distinguish chemical phenomena.*—It is easy, then, to know whether any event among those we see in nature or in the arts, is to be explained by the principles of Chemistry. In the flow of water, the turning of a wheel, the buzzing of a saw: in the motion of wind, the rising of vapors, and the return of the rain, there are no changes taking place in the nature of bodies; with these events, then, Chemistry has nothing to do. But in the burning of wood, the ripening of fruits, the decay of plants; in the process of getting metals from the ores; in the manufacture of paper from rags or straw, we may at once see that there are changes occurring in the nature of substances, and the explanation of all such phenomena is to be given by the science of Chemistry.

We now see clearly that the work before the chemist is to study the properties of matter, more particularly those which it can not manifest except by a change in its nature; to learn the composition of matter, and, as far as may be, the explanations of those phenomena in which there are changes taking place in the nature of bodies. But the life time of an individual would be scarcely long enough to complete so great a labor. In the brief time given to this study in a student's course, let us endeavor to become familiar only with the simplest or introductory principles of the science.

CHAPTER I.

OF THE COMPOSITION OF BODIES.

General Statement.—All substances are either elements, compounds, or mixtures.

(3.) Elements are substances which have never been decomposed. We may notice in the outset nitrogen, oxygen, hydrogen, and carbon.

I.—NITROGEN.

A.—Nitrogen may be obtained from the air for examination. It is a colorless gas, which extinguishes fire and will not support life.

1. *Obtained from air.*—There are large quantities of nitrogen in the atmosphere, but there are large quantities of oxygen with it. By burning phosphorus in a portion of air the oxygen will be taken away and the nitrogen left. For this purpose let a piece of phosphorus the size of a large pea be placed on a cork floating upon the water of a cistern. Touch it with a hot iron, and quickly invert over it a gallon jar. (See Fig. 1.) The phosphorus burns with a beautiful light, while milk-white vapors fill the jar. These vapors will be

gradually absorbed by the water which will rise into the jar. The space above the water at last is filled with nitrogen.

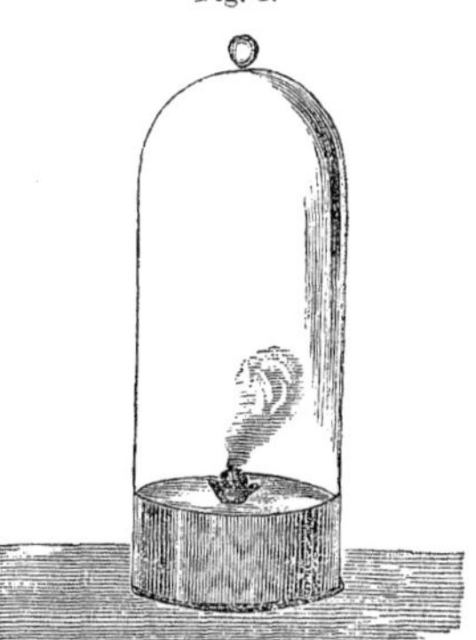
Fig. 1.

2. *Its physical properties.*—The nitrogen is now seen to be a gas, perfectly colorless and transparent. It is without odor or taste and a little lighter than air, its specific gravity being .972.

3. *Its chemical properties.*—If a lighted taper be lowered into a jar of nitrogen it will be extinguished as quickly as if plunged into water. The gas will neither burn nor allow other things to burn in it. So, too, if an animal were put into this gas, death would soon follow. It is not poisonous, but kills simply because it has no power to support life. A bandage over the face which would shut all air away from the lungs would kill in the same way.

4. *It occurs in nature.*—Nitrogen is a very abundant element. It forms about four-fifths by weight of all the atmosphere, and besides this it is an important constituent in many animal and vegetable bodies.

II.—OXYGEN.

B.—Oxygen may be obtained in many ways. It is usually prepared by heating potassic chlorate. It is a colorless gas, in which bodies burn better than in air. Animals can not live without it. It is the most abundant element in nature.

1. *Obtained from potassic chlorate.*—Potassic chlorate (chlorate of potash) is a white solid, about 39.2 per cent. of its weight being oxygen. To set the oxygen free the chlorate must be heated. For this purpose it is finely powdered, mixed with about an equal weight of black oxide of manganese, and put into a flask F, Fig. 2. A bent tube reaches through the cork of the

Fig. 2.

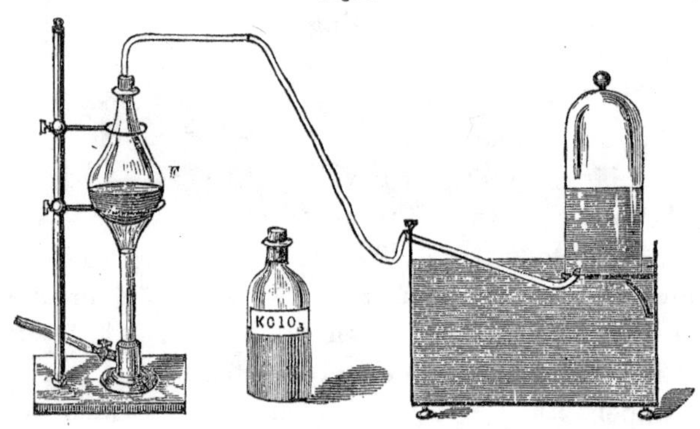

flask and over into the water of the cistern, and upon the shelf of the cistern there stands an inverted jar filled with water. Now, when the flask is heated, the chlorate will be decomposed, and oxygen being set free will pass through the bent tube and bubble out of the water. If the end of the tube is brought under the mouth of the jar the oxygen will rise into the jar, which in a little time will be filled.

2. *Its physical properties.*—The oxygen is a colorless and transparent gas; without odor or taste; a little heavier than air, its specific gravity being 1.1056.

3. *Its chemical properties.*—A lighted taper lowered into a jar of oxygen burns with surprising brilliancy. A glowing spark upon the wick is all that is needed:

the oxygen instantly and with a slight explosion, kindles it into a vivid flame. Experiments like these illustrate the fact that bodies which burn in air will burn with greater vigor in oxygen.

Again: let an iron wire or a steel watch spring be tipped with a bit of wood. Set fire to the wood and at once plunge it into a jar of oxygen. Quickly the metal takes fire and burns--the iron with a steady and beautiful light, or the steel with a multitude of star-like sparks. Experiments such as this illustrate the fact that substances which do not burn in air may burn with great rapidity in oxygen.

It not only supports combustion, it is also necessary to the life of animals. It is in the air, and animals breathe it; it goes into their blood and purifies it. And yet were an animal to breathe pure oxygen, death would surely follow. By being mixed with nitrogen its violent action is toned down so that the most delicate organ may not only withstand it, but be invigorated by its presence.

Allotropism is a chemical property worthy of especial notice. It is the property in virtue of which the same element under different conditions may show different properties. To illustrate the allotropism of oxygen, let strips of unsized paper be soaked in a solution of potassic iodide (iodide of potassium) mixed with starch. If these strips be hung in a jar of air no action or change will occur; but if a few drops of ether be added, and a hot glass rod be put into the jar (Fig. 3), the paper will very soon become colored blue. The oxygen, which at first could not attack the iodide, has been changed by the ether and the heat so that it can. This allotropic form of oxygen is called *ozone*.

That ozone is nothing but oxygen may be proved. By passing electric sparks thro' pure oxygen ozone is formed. Now, electricity is not a kind of matter, and hence can add nothing to, nor can it take any thing from, the element oxygen. Ozone is therefore but another form of oxygen. Moreover, if left alone in the jar, ozone will in time return to the form of oxygen.

Fig. 8.

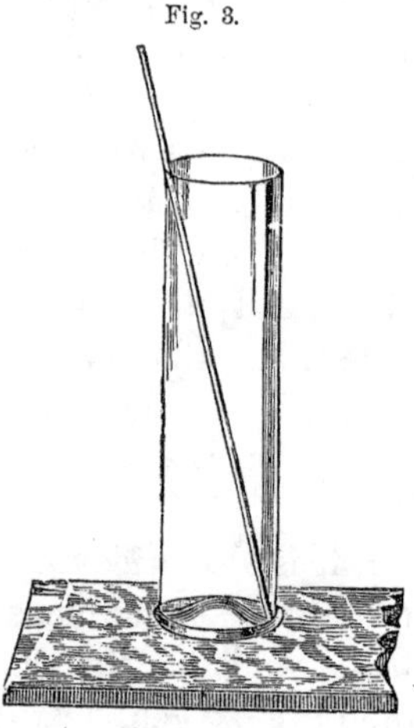

Vigorous as is the chemical action of oxygen, it is still more vigorous in the form of ozone. This is the chief difference in the action of these allotropic forms.

4. *It is abundant in nature.*—Oxygen is the most abundant element in nature: minerals, plants, and animals alike contain large quantities of it. One-fifth part of the air by weight is oxygen, eight-ninths of all the water on the globe, and about one-half of all the solid rocks. Besides this it forms about four-fifths of the weight of vegetable bodies, and about three-fourths of that of animals. It is, perhaps, not too much to say, that one-half of all the matter of the world, as far as it has been examined, is oxygen. And yet when freed from its prisons in solid and liquid bodies, oxygen is a gas, invisible as air and but little heavier.

III.—HYDROGEN.

C.—Hydrogen may be obtained from water. It is a

very light and colorless gas; burning with an almost colorless flame.

1. *Obtained from water.*—Pure water consists of hydrogen and oxygen, and there are many ways in which they may be separated. Many of the metals have power to take the oxygen from water. If a piece of sodium be dropped upon water it melts into a globule, floats, and runs briskly around over the surface, taking the oxygen to itself and letting the hydrogen go free. If the sodium be confined in a net of wire gauze it may be brought under the mouth of a test tube previously filled with water. (Fig. 4.) The hydrogen will then be collected in the tube.

Fig. 4.

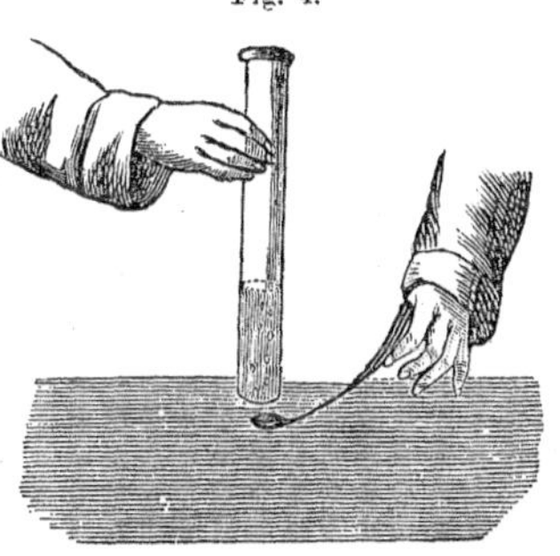

A more practical way of getting hydrogen, is to decompose water by means of zinc and sulphuric acid.

Into a bottle, B, Fig. 5, are put some fragments of zinc. Through the well-fitting cork pass two tubes, one a funnel-shaped tube reaching down into the bottle, the other a bent tube reaching over to the cistern of water. If now a mixture of water and sulphuric acid is poured through the funnel until the lower end of its tube is covered, hydrogen will flow rapidly through the bent tube, and may be collected in jars upon the shelf of the cistern.

2. *Its physical properties.*—Hydrogen is a colorless gas, and when pure has neither taste nor odor. It is the lightest of known substances, its specific gravity

being only .0692. On this account, among other reasons, chemists are inclined to make it the standard for specific gravity. Calling its specific gravity 1, that of oxygen is 16, and that of nitrogen is 14.

3. *Its chemical properties.*—Hydrogen is a combustible gas. If a small jar, filled with it, is carefully lifted from the cistern, and a lighted taper is quickly pushed up into it, the gas takes fire with a slight explosion. Let a bottle containing zinc, water, and sulphuric acid be provided with a jet pipe passing through a well-fitting cork. After the hydrogen has driven out all the air, if the stream of gas is touched with a lighted taper, it will take fire and continue to burn with a steady flame. Notice how nearly colorless is this flame; but when a small wire is put into it, you may see how quickly it will glow with heat. The flame of burning hydrogen gives but little light, but it is the source of intense heat.

Fig. 5.

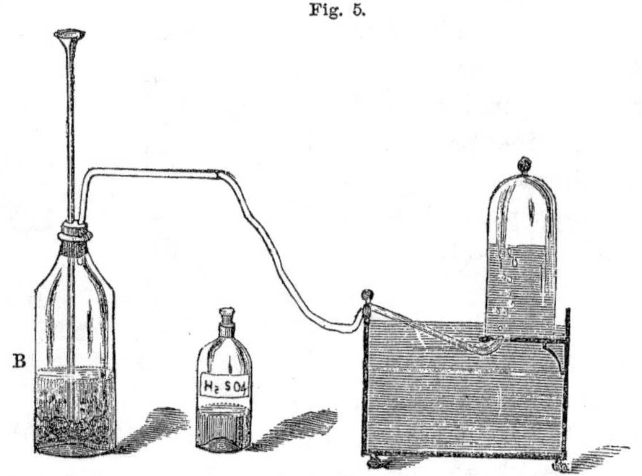

When mixed with air, hydrogen explodes; if with oxygen in the proportions of 2 to 1 by volume, the explosion is deafening. If it takes place in free air it is not dangerous, but if the mixture is confined in a tight bottle the fragments of the bottle will be driven with violence as by a charge of gunpowder.

IV.—CARBON.

D.—Carbon, unlike the elements just described, is a solid substance. It may be obtained by heating wood in a close vessel, or by burning it in a limited supply of air. It occurs in three allotropic forms: diamond, graphite, and coal.

1. *Obtained from wood.*—When splinters of wood are heated in a test-tube closed with a cork through which passes a small tube, considerable quantities of vapor and gas escape, and the wood turns black. The black mass that remains when the gas is no longer given off is charcoal, one form of carbon. In the same way, on a large scale, wood is put into retorts and heated intensely; its liquid and gaseous constituents are driven off, while the solid carbon remains.

Or again: if we light one end of a splinter of wood, and slowly push the burning part into the mouth of a test-tube, the part in the tube will only glimmer in the small supply of air, or be extinguished altogether. The black residue is carbon. In the same way when charcoal is required in large quantities, piles of wood are burned under a covering of sod and moist earth, where but little air can reach them. Smoldering sometimes for weeks, the gaseous parts of the wood are at last all driven off, while the carbon, keeping the form of the original sticks, with knots and annual rings still perfect, all, however, much reduced in size and weight, remains.

2. *It is found in nature.*—All the different varieties of coal are but different impure forms of carbon, the remains of a vegetation which grew ages before the period of man's creation.

Enormous beds of limestone are found on every continent, but not a molecule of limestone occurs which does not contain an atom of carbon.

And more: not a single organized body, from the lowest form of plant to the highest form of animal exists, in which carbon is not an important element.

3. *The diamond.*—Of this most valuable gem little need be said. Its great power to refract light and its wonderful hardness are familiar: these are the properties which render the gem so valuable in the arts. The light flashing from the different sides of the crystal makes it the most brilliant of ornaments: its extreme hardness renders it valuable in the construction of pivots in delicate instruments where friction is to be avoided.

A roughly rounded pebble found in the sand and clay of India or Brazil, the diamond would hardly be picked up by one whose eye had not been trained to recognize it; but when polished it is at once the most beautiful and the most indestructible of gems. It can neither be melted nor dissolved: it may, however, be burned when heated intensely in oxygen gas.

4. *Graphite.*—This substance, also called plumbago, and more familiarly known as black-lead, is taken from the earth in large quantities for the manufacture of lead pencils. It has a black and shining luster, and it is a good conductor of electricity, being in these respects much like metals.

5. *Allotropic forms.*—Charcoal, the diamond, and graphite, are but different forms of the element carbon.

They differ in hardness, in color, in weight, and in many other physical properties. They are alike infusible, alike able to resist the action of substances which attack most other bodies, alike in being combustible, and alike in yielding the same substance (carbonic di-oxide) when burned. That they are also alike in being of vegetable origin is believed, but of the mode in which the diamond and graphite have been made from vegetable matter little, if any thing, is known.

(4.) Only 63 elements have yet been discovered. Of this number 14 are called non-metals: the remaining 49 are metals. All known forms of matter are thought to be made of these elements.

1. *The 63 elements.*—Some of these substances are so very rare as to be of little importance to the chemist; even the existence of two or three is still in some doubt. The following table contains a complete list of their names, together with other matter for future reference.

NON-METALS.

Names.	Symbols.	Quantivalence.	Combining weight.	Combining volume.
Boron	B	III	11	
Bromine	Br	I	80	1
Carbon	C	IV	12	
Chlorine	Cl	I	35.5	1
Fluorine	F	I	19	
Hydrogen	H	I	1	1
Iodine	I	I	127	1
Nitrogen	N	III	14	1
Oxygen	O	II	16	1
Phosphorus	P	III	31	½

Names.	Symbols.	Quantivalence.	Combining weight.	Combining volume.
Selenium	Se	II	79.5	1
Silicon	Si	IV	28	
Sulphur	S	II	32	1
Tellurium	Te	II	129	

METALS.

Names.	Symbols.	Quantivalence.	Combining weight.	Combining volume.
Aluminum	Al	II	27.4	
Antimony	Sb	III	122	
Arsenic	As	III	75	½
Barium	Ba	II	137	
Bismuth	Bi	III	210	
Cadmium	Cd	II	112	2
Cæsium	Cs	I	133	
Calcium	Ca	II	40	
Cerium	Ce	II	92	
Chromium	Cr	II	52.2	
Cobalt	Co	II	58.7	
Copper	Cu	II	63.5	
Didymium	D	II	95	
Erbium	E		112.6	
Glucinium	Gl		9.3	
Gold	Au	III	197	
Indium	In	II	74	
Iridium	Ir	IV	198	
Iron	Fe	II	56	
Lauthanum	La	II	92	
Lead	Pb	II	207	
Lithium	Li	I	7	
Magnesium	Mg	II	24	
Manganese	Mn	II	55	
Mercury	Hg	II	200	2

Names.	Symbols.	Quantivalence.	Combining weight.	Combining volume.
Molybdenum	Mo	VI	96	
Nickel	Ni	II	58.7	
Niobium	Nb	IV	94	
Osmium	Os	IV	199.2	
Palladium	Pd	II	106.6	
Platinum	Pt	IV	197.5	
Potassium	K	I	39.1	
Rhodium	Rh	II	104.4	
Rubidium	Rb	I	85.4	
Ruthenium	Ru	IV	104.4	
Silver	Ag	I	108	
Sodium	Na	I	23	
Strontium	Sr	II	87.5	
Tantalum	Ta	V	182	
Thallium	Tl	I	204	
Thorium	Th	IV	115.7	
Tin	Sn	IV	118	
Titanium	Ti	IV	50	
Tungsten	W	VI	184	
Uranium	U	III	120	
Vanadium	V	III	51.3	
Yttrium	Y	II	61.6	
Zinc	Zn	II	65.2	
Zirconium	Zr	IV	89.6	

(5.) The non-metals lately discovered have been named from some characteristic property they possess: the names of metals in addition to this receive the common termination um. These names are represented by symbols.

1. *Naming the elements.*—Many of the elements re-

tain the names given them before any rules of nomenclature were fixed. Such are sulphur, iron, and the precious metals.

In the case of non-metals lately discovered, some property of the element is seized upon to furnish a name. Thus, *chlorine* was seen to be a greenish gas: the name, chlorine, is from the Greek word meaning green. *Hydrogen* when burned was found to produce water, and its name is from the Greek words meaning, to form water. To distinguish the metals from non-metals their names end in *um*. Potassium, sodium, and calcium are examples.

2. *Symbols.*—Instead of writing the names of elements in full, chemists have agreed to use a set of symbols to represent them. These symbols are the initial letters of the names (often of the Latin names), or in case two elements have the same initial, a small letter is added to the initial. Thus O stands for oxygen; N, for nitrogen; H, for hydrogen. Of carbon, cobalt, and copper (Latin cuprum), the symbols are C, Co, and Cu. The symbols of all the elements are given in the table preceding.

(6.) Compounds are substances made of two or more elements, but whose properties differ from those of the elements which compose them. We may notice water and carbonic di-oxide, as examples.

1. *Compounds.*—Iron left in moist air is soon covered with rust; indeed, the whole piece may be finally changed to iron rust. Now, this rust is made of the metal iron, and the non-metal oxygen which the iron has taken from the moisture of the air. But how dif-

ferent are the qualities of rust from those of either iron or oxygen! It is a *compound* of these two elements.

Other metals may be rusted by being exposed to the air, or, if need be, at the same time heated. Mercury, for example, if heated for a long time in air forms a rust composed of mercury and oxygen. It is a red powder in which both these substances exist, but which show the properties of neither. That this "red oxide of mercury" contains both elements may be easily proved. For this purpose a tube containing a small quantity of the oxide is tightly closed with a perforated cork from which a bent tube passes over to the water of the cistern. (Fig. 6.) By heating the tube the air will be driven out, and if afterward a small vial filled with water is inverted over the end of the tube, it will be quickly filled with a colorless gas.

Fig. 6.

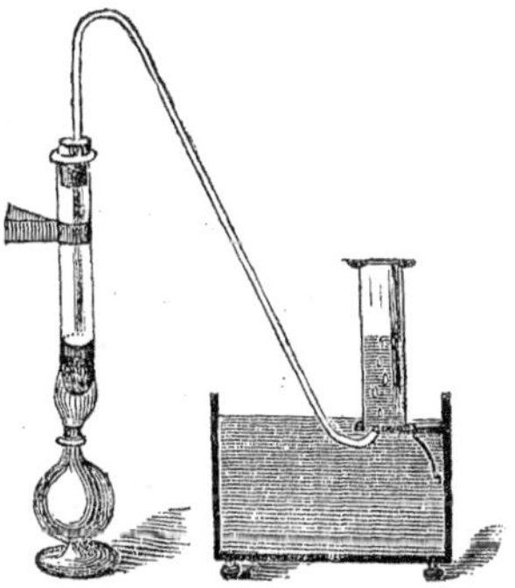

Little globules of shining mercury will be found clinging to the sides of the tube, while the gas, tested with a lighted taper, is found to be oxygen.

Very few of the sixty-three elements are found as elements in nature: they are for the most part known only after being set free from the compounds in which they occur. Nearly all substances, then, are compounds or mixtures of compounds.

I.—WATER.

A.—Pure water is a compound of oxygen and hydrogen. In nature it occurs in the solid form as ice, in the liquid state as water, and in the gaseous form as steam.

1. *Water is a compound.*—We have seen that hydrogen is a combustible gas. But when burnt it is no longer hydrogen: what change takes place? Let the following experiment teach us:

Over the flame of burning hydrogen hold a jar filled with oxygen as shown in Fig. 7. Instantly the sides of the jar, a moment before thoroughly dry and

Fig. 7.

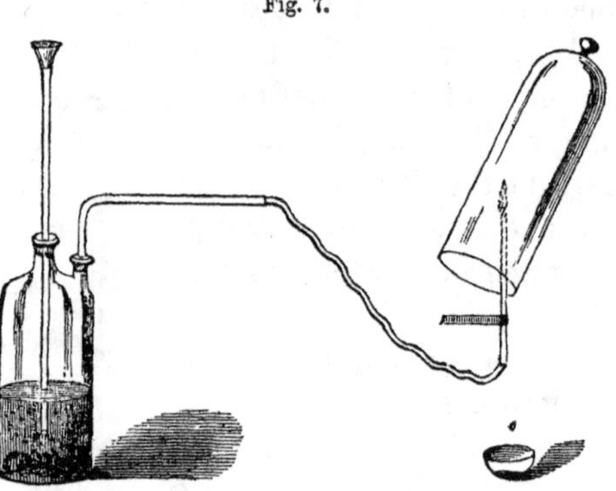

clean, are dimmed with steam, and if they are kept cold, the steam is condensed until it trickles in delicate streams of water down to the edge, and finally drops from the jar. Now, the hydrogen from the bottle and the oxygen in the jar must have formed this water; no other substances took any part in the flame. Hydrogen, a combustible gas, and oxygen another gas, which more than any other quickens flame, have formed the water which quenches fire. Water is thus shown to be a compound of hydrogen and oxygen.

2. *Its properties.*—Water is too familiar to need a

long description. Let us notice only that it evaporates at all temperatures, keeping the air charged with its invisible vapor, which when condensed produces fogs and mists, dews and rain ; that it boils at 100°C. (212° F.), freezes at 0° C. (32° F.), and has its greatest density at 4° C. (39° F.). At a temperature of 15° C. (59° F.) it is 819 times heavier than air. Of the chemical properties of water we shall learn in future lessons.

II.—CARBONIC DI-OXIDE.

B.—Carbonic di-oxide is a compound of carbon and oxygen. It may be obtained from marble by the action of hydrochloric acid. It is a heavy colorless gas which extinguishes fire, and when breathed causes death.

1. *Carbonic di-oxide a compound.*—Into a jar of pure oxygen, hang by means of a wire a piece of charcoal bark on which there is a spark of fire (Fig. 8). Quickly the charcoal bursts into a beautiful and vigorous combustion. When the burning is over, we may see that either the whole or a part of the charcoal has disappeared, and that the jar is still full of colorless gas. But if lime water is poured into the jar, the gas colors it a milky white. Now, oxygen can not turn lime water milky, hence another colorless gas must have been formed. This new gas is carbonic dioxide (carbonic acid) ; and since carbon and oxygen are the only substances which took any part in the experi-

Fig. 8.

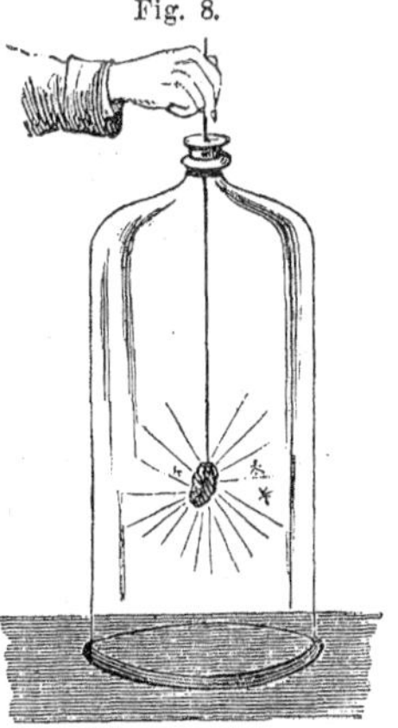

ment, carbonic acid must be composed of these elements.

2. *Obtained from marble.*—Carbonic di-oxide is a constituent of marble, and may be obtained from it by the following process. Small fragments of marble are put into a bottle whose cork is provided with two tubes, one reaching to the cistern, the other a funnel tube reaching nearly to the bottom of the bottle. (Fig. 9.) Water is poured through the funnel until the

Fig. 9.

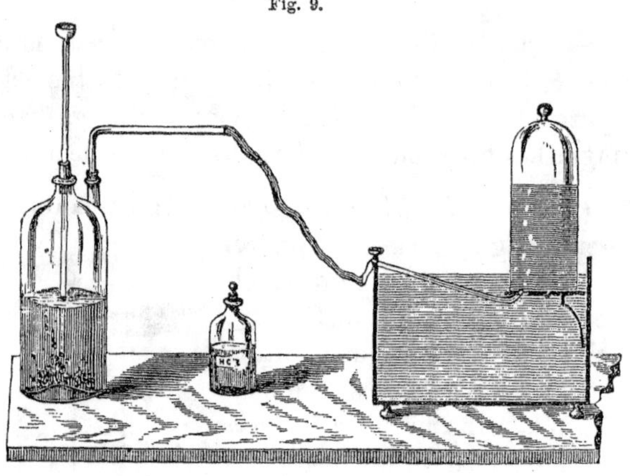

lower end of its tube is covered, and hydrochloric acid is then added. A violent agitation quickly begins in the bottle; carbonic di-oxide is set free, and is collected in the jar over the water.

3. *Its properties.*—Carbonic di-oxide is a colorless gas: in this respect it is like the others which have been examined. If a lighted taper is plunged into it, the flame is instantly extinguished; in this respect carbonic di-oxide resembles nitrogen. If lime water is exposed to its action it will be turned milky: this

effect can not be produced by either of the other gases.

Carbonic di-oxide is much heavier than air; its density being 1.529. To illustrate: let the end of the bent tube from the bottle in which this gas is made be placed in a jar, standing with its open mouth upward. In a little time a lighted taper, lowered into the jar, is quenched, showing that the jar is full of the gas. The gas remains in the open vessel as water would, and just for the same reason, it is heavier than air. Indeed, it may be poured from one vessel to another like water: its flow can not be seen, but a lighted taper or lime water will show its presence in the second vessel and its absence from the first.

Unlike other gases thus far examined, carbonic di-oxide may be condensed to a liquid and even to a solid state. As steam when cooled becomes water, so this gas when made cold enough becomes a liquid; the temperature required is $-106°$. And as, by being cooled still farther water is frozen, so carbonic acid may be changed from the liquid to the solid form. To reduce the temperature low enough, the liquid acid is suddenly exposed to air at ordinary temperature: it evaporates so swiftly that the large amount of heat, taken up by the change from the liquid to the gaseous form, leaves the rest of the liquid so cold that it freezes. Gases may be liquefied not only by cooling them, the same effect may be produced by pressure. Thus steam whose temperature is kept up to 100° C. will be changed to water by pressure: so carbonic di-oxide, by a greater pressure, may be liquefied. At a temperature of 0° C. it takes a pressure of 36 atmospheres, equal to 540 lbs. to the square inch.

Carbonic di-oxide, when breathed, poisons the system. If pure, it causes spasms in the air-passages, which may prevent its going to the lungs; but when mixed with air, it can be easily breathed, and, if in sufficient quantities, it causes death. Even as much as one-tenth per cent. is injurious.

4. *It occurs in nature.*—In small quantities carbonic di-oxide is always present in air and water: in the solid earth, being a constituent of all limestone rocks, its quantity is immense.

(7.) The composition of a compound may be determined by analysis and synthesis. There are various methods of analysis: we may notice, 1st, by electricity; 2d, by the prism; and 3d, by the chemical action of substances upon one another.

1. *Analysis.*—Any process by which a compound may be separated into its constituents, and its composition determined, is called analysis. If in the process only the names of the constituents are found, the analysis is *qualitative;* if their proportions are found, the analysis is *quantitative.*

2. *By electricity.*—Many substances may be decomposed, and their composition found, by the action of electricity. Such a process is called *electrolysis.* The most interesting example of electrolysis is that of water, and this will be the only example we need to notice now. Two platinum strips are inserted in a jar of water, and over them are inverted two long and slender tubes, previously filled with water. (See Fig. 10.)

The wires of a galvanic battery are then inserted in the screw cups, *s s.* Instantly a torrent of gas bubbles rises in each tube, and will continue to do so until the tubes are filled. These two gases are the constituents of water. As the experiment goes on, it will be noticed that one tube is being filled faster than the other; in-

Fig. 10.

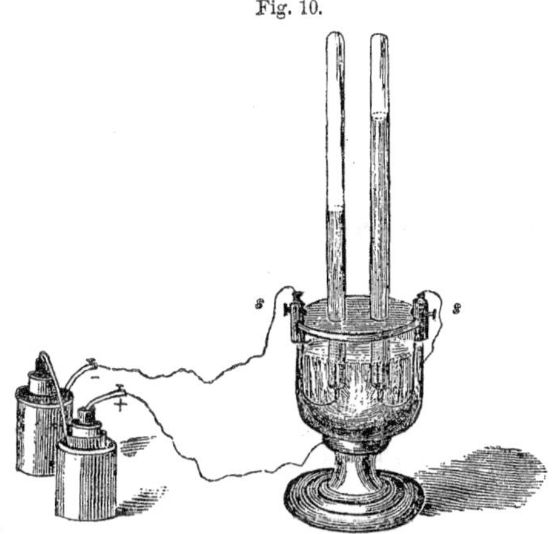

deed, being of equal size, one fills just *twice as fast* as the other. If the gases are tested, that which is most rapidly set free is found to be hydrogen, the other oxygen. The experiment teaches that water is composed of hydrogen and oxygen, in the proportion by volume of 2 of hydrogen to 1 of oxygen.

The oxygen will invariably rise over the positive electrode of the battery: it is therefore called the electro-negative element, the hydrogen is electro-positive. Many other liquids may be analyzed in a similar way, and their constituents are electro-negative or electro-positive, according to the electrode over which they appear.

3. *By the prism.*—When the light of a burning substance is passed through a prism it is decomposed, and

the appearance of the spectrum will depend upon the nature of the burning substance. Hence the constituents of a compound may be told by the appearance of its spectrum. This method of detecting the presence of substances is called *spectrum analysis.*

For example: When the spectrum of burning sodium is viewed through a telescope, two bright yellow lines, very close together and of surprising brilliancy, are seen in the yellow part of the spectrum. Or, if potassium is burned, a single crimson line and another of blue, both of great beauty, will be seen in opposite ends of the spectrum. Nor will the appearance of either set be changed by the presence of the other; for, if a mixture of sodium and potassium is burned, the observer will see both sets in the same spectrum, their place, size, and brightness the same as when each was formed alone. The spectra of many elements have been carefully studied; they can be easily recognized by one who has previously made their acquaintance.

This method of analysis has already led to the discovery of four new elements; their names are: Rubidium, Cæsium, Thallium, and Indium.

4. *By chemical action.*—The methods of analysis just described are limited in application. The general method is by the chemical action of bodies upon each other. More than a hint of this method would be out of place here. But suppose, for example, that a solution of unknown substances in water is to be analyzed. A few drops of hydrochloric acid may be added. If a white, solid substance is formed, it shows the presence of either silver, mercury, or lead. To find out which of these metals is present, a little ammonia is added: if the solid disappears again, the metal is silver; if it

simply turns black, mercury is present; but if it remains unchanged, the metal is lead. If the acid fails to produce the white solid, or *precipitate* as any solid formed in this way from solution by the action of chemicals is called, then these three metals are counted absent, and some other agent is used by which another group of metals may be detected.

5. *Synthesis.*—The composition of a compound may be found by causing its constituents to combine, and noticing the proportions required. This process is called synthesis. The synthesis of water may illustrate: it may be made in an apparatus called a *eudiometer*. This instrument is a glass tube, open at one end and carefully graduated, having two metallic wires passing through the glass near the closed end, and almost touching each other inside. To use it, measured quantities of hydrogen and oxygen, say 50 volumes of each, are put into it. The tube should still be at least half full of water, and should be firmly held with its open end in the cistern. If now electricity from a Leyden jar is passed through the mixture (see Fig. 11), the two gases combine with violent explosion; water will be formed by their union, and water from the cistern will rise into the tube to fill the space they occupied. The quantity of gas left will be found to be 25 volumes, and, if tested, will be found to be oxygen. It is evident that 50 volumes of hydrogen have taken 25 volumes of oxygen to form water, and the experiment

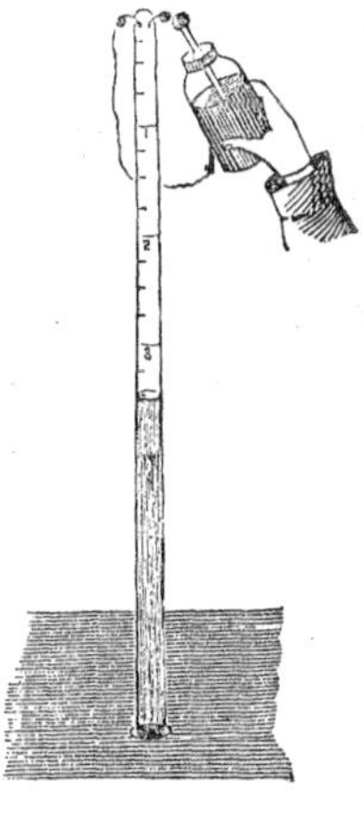

Fig. 11.

teaches that water is composed of hydrogen and cxygen in the proportions of two volumes of hydrogen to one of oxygen.

(8.) Mixtures are substances made up of two or more elements or compounds, each of which retains its own properties. As examples we may notice air, and water as it is found in nature.

1. *Mixtures.*—When two or more substances are put together without causing any new properties to arise, the substance formed is called a mixture. Sugar and salt may be crushed to the finest powder and thoroughly shaken or ground together, yet no change will be produced in their character. The mass has no property at last not possessed by one or the other at first; it is not a compound but simply a mixture.

Now, such is the character of almost every body found in nature. If elements are seldom found uncombined, so both elements and compounds are still more rarely found in nature pure. Rocks and soils are mixtures. The air we breathe is a mixture of elements and compounds, and it is no exaggeration to say that not a drop of pure water exists upon the earth.

I.—AIR.

A.—Air is a mixture of nitrogen, oxygen, carbonic di-oxide, and water, with small quantities of many other gases.

1. *Nitrogen is a constituent of air.*—Let a jet of hydrogen be burned in a bottle of air. For this purpose an india-rubber tube from the bottle B, Fig. 12, ends in a jet pipe which passes tightly through one of

two holes in a cork which fits the air bottle A. Set fire to the jet of hydrogen and insert it in the air bottle, whose neck may then be lowered into water. The hy-

Fig. 12.

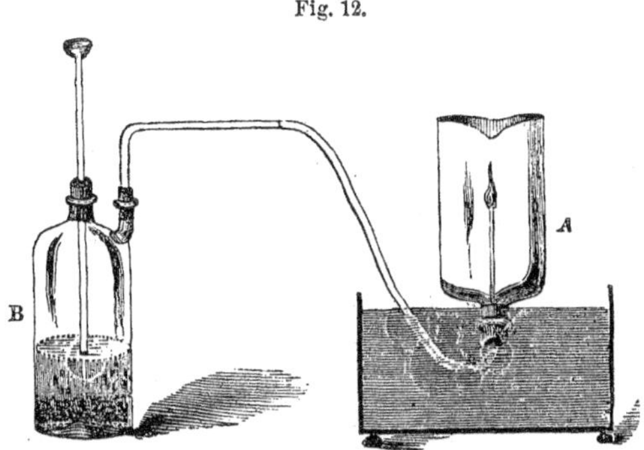

drogen burns for awhile and then ceases; the water in the mean time rising a little distance into the bottle.

If the jet pipe be now removed and the remaining gas tested, it will be found to be nitrogen. Hence nitrogen is a constituent of the air.

2. *Oxygen is a constituent of air.*—We have seen that mercury, when heated for a long time in air, is changed to a red oxide of mercury, and that if this oxide is heated (see p. 25) the metallic mercury will be restored while oxygen will be given off. Now observe: the oxygen set free from the oxide in the last heating must have been taken from the air in the first. The experiment teaches that oxygen is a constituent of air.

3. *Carbonic di-oxide is a constituent of air.*—Let a goblet of lime water stand for a few hours exposed to the air: a crust will be formed on its surface. This crust is of the same material that rendered lime water white when acted upon by carbonic di-oxide (see p. 28),

and we therefore infer that carbonic di-oxide is a constituent of the air.

4. *Water is a constituent of air.*—That a vessel of ice water on a warm day will in a little time be covered with drops of dew, is a fact sufficiently familiar. Now, the water collected on the vessel can have come only from the air. It must have been in the air in the form of invisible vapor, which, cooled by the cold sides of the vessel, is condensed. Hence water is a constituent of the air.

5. *The proportions of nitrogen and oxygen.*—By far the larger part of the atmosphere consists of nitrogen and oxygen; their proportions may be found by experiment. A vessel, V, Fig. 13, has two openings closed with stop-cocks, *c* and *d*. To the upper stop-cock is attached

Fig. 13.

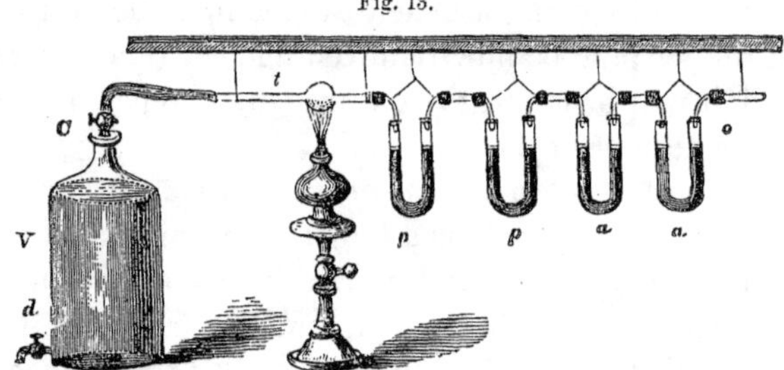

a series of tubes, one, *t*, with a bulb in which is put some copper to be heated by a lamp below; others, *p p*, containing potash, and others still, *a a*, containing sulphuric acid.

The capacity of the vessel is accurately known, and at the beginning of the experiment it is quite full of water. When the stop-cocks are opened, air will flow through the series of tubes entering at *o*, until the vessel is filled. In passing through *a a*, all its water will be taken out by the acid; in going through *p p*, all its carbonic di-oxide (carbonic acid) will be taken by the potash; in passing over the heated copper in the bulb, it will give up all its oxygen, and the nitrogen will be left to enter the globe alone. Now notice: if the tube *t* be accurately weighed before and after the experiment, what it has gained will be the weight of the oxygen taken from the air that passed through it. The vessel V is filled with the nitrogen from the same air, and, since the capacity of the vessel is known, the quantity of gas is also known. By this means it has been found that air, from which water and carbonic di-oxide have been taken, contains, by weight, 76.9 parts nitrogen to 23.1 parts of oxygen in 100 parts of air.

6. *The proportions of water and carbonic di-oxide.*—Now the quantities of water and carbonic di-oxide might also be found in this experiment by weighing the tubes in which they are left, were it not that in so small a quantity as the vessel full of air they are too small to be very appreciable. If, however, larger quantities of air are passed, the increase in weight will be quite enough. It has been found that the quantities of these substances vary at different times and places, being always very small.

The quantity of water is exceedingly variable. It depends upon locality and temperature. At 15° C. the largest quantity that air can hold is $\frac{1}{80}$ of its own weight. Commonly the amount is far less than this.

The proportions of carbonic di-oxide is usually given by volume. It varies from $\frac{1}{5000}$ to more than $\frac{1}{1000}$, averaging about $\frac{1}{2500}$.

7. *The air is a mixture.*—In the properties of the atmosphere we discover none that do not belong to one or the other of its constituents. The oxygen causes bodies to burn, not so freely, of course, as if it were pure. The nitrogen, if pure, would quench all flames ; in the air it hinders the burning which it can not quench. So, too, the water vapor, by being cooled, is condensed into clouds and rain water, just as the pure substance first becomes visible as white vapor and then changes to water. And finally, the carbonic di-oxide of the air turns lime water white just as the pure gas would do, only in a less degree. Hence the air is a mixture of its constituents, not a compound.

II.—NATURAL WATERS.

B.—The solvent power of water is greater than that of any other liquid; on this account pure water is never found in nature. All natural waters are mixtures.

1. *The solvent power of water.*—A substance is said to be dissolved in water when its particles are so completely separated and scattered through the liquid as to be invisible. Its cohesion is entirely overcome by the stronger force of adhesion between its particles and those of water. A substance that may thus disappear is said to be *soluble*, and the fluid which contains it is called a *solution*, while that which dissolves it is called a *solvent.* Salt is soluble in water ; the brine is a solution, the water a solvent.

But it is well known that only a certain quantity of salt can be dissolved in water, all added beyond that will remain undissolved. It is so with other solids;

water can only dissolve a limited quantity. A solution in which the fluid has all it can hold of a soluble body is said to be *saturated.*

The solvent power of water may be increased by heat. To this, however, there are some exceptions. Common salt, for example, will dissolve about equally in hot and in cold water, and lime much better in cold than in hot water.

Many gases, also, are soluble in water. Oxygen and nitrogen are examples, small quantities of both these gases, taken from the air, may be found in all natural waters. Like solids, each gas has its own degree of solubility; oxygen for example is more soluble than nitrogen, while carbonic di-oxide is much more soluble than either. Of this gas, water will dissolve about its own volume. Others are still more soluble; of ammonia gas, water at 0° C. will dissolve 1,049 times its volume.

Other liquids beside water are often used as solvents, but none may be so universally applied.

2. *Natural waters are impure.*—Because water can dissolve so many other bodies we may not expect to find pure water upon the earth. Coming from the clouds it dissolves the gases of the atmosphere; sinking into the soil, it dissolves the minerals it meets, and flowing over rocks, it dissolves their materials. Hence the waters of all seas, lakes, rivers, and wells, are, without exception, impure, the kind and quantity of their impurities depending on the material over which they have passed.

3. *Mineral springs.*—Salt springs are those whose waters have dissolved out the salt from the soil through which they have flowed. In another place some other

substance may exist in the soil or rocks, and the water, dissolving these, forms other kinds of mineral springs.

4. *The saltness of the sea.*—In a similar way we may account for the saltness of the sea. Common salt is scattered in small quantities through most soils and is dissolved by water which trickles through them. This water collects in rivers and finds its way to the sea. From the sea, water can only escape by evaporation; by this process, only pure water escapes; the salt is left behind. Age after age this work goes on, and large quantities of salt have thus already been accumulated in the sea.

Salt lakes are found, such as the great salt lake of Utah. They are lakes without an outlet, so that while the water may escape by evaporation, there is no escape for the salt.

5. *Such solid impurities increase the weight of the water.*—A very pretty experiment shows that brine is heavier than fresh water. In the first place a small vial is partly filled with a colored liquid, so that when corked it will just sink in fresh water. Into a tall jar, Fig. 14, put fresh water several inches deep, and into this the little vial. Now by means of a long funnel-tube a saturated solution of salt, may, with care, be poured to the bottom of the jar without mixing with the fresh water above. The fresh water will be lifted, floating on the brine,

Fig. 14.

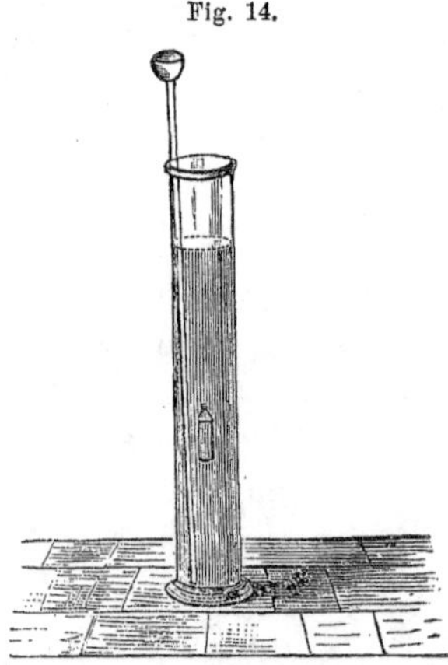

while the little vial, also lifted, marks the dividing line between them.

III.—DIFFUSION AND OSMOSE.

C.—The adhesive attraction between molecules of different liquids will often cause them to mix if they are merely brought in contact. This action is called the diffusion of liquids. Diffusion may take place through membranes and other porous solids: it is then called osmose. Gases also mix by diffusion and osmose.

1. *The diffusion of liquids.*—If in a tall jar some colored water is placed, alcohol may, with care, be poured upon it without mixing the fluids. Being lighter than water the alcohol floats, and being colorless, the line of division may be distinctly seen. After a few hours the liquids in the jar will be found colored uniformly throughout. The heavy water has risen; the lighter alcohol has sunk, and a uniform mixture is formed. Liquids which when shaken together will remain mixed, will diffuse when brought in contact, but those which separate again when standing awhile will not diffuse.

Fig. 15.

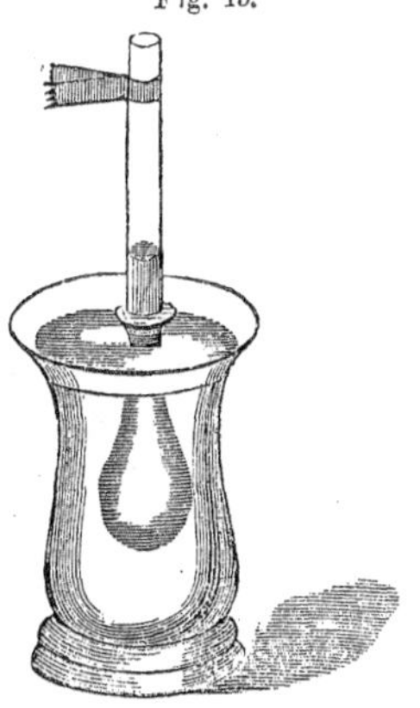

2. *Osmose of liquids.*—A thin membrane or porous substance will not prevent liquids from mixing. To illustrate: let a bladder be firmly tied over the end of a lamp-chimney, or other glass tube; and, being filled with alcohol, let it be pressed a little way into

a vessel of water (Fig. 15). After a time, the fluid will be seen gradually rising in the chimney. The water flows in to mix with the alcohol, and a smaller quantity of alcohol flows out at the same time. This mixing of liquids through porous solids is called *osmose.*

3. *The diffusion of gases.*—Diffusion among gases is more rapid than among liquids. The following experiment will clearly illustrate this important property. Two strong bottles with narrow necks, one filled with oxygen, the other with hydrogen are placed with their open necks together, the oxygen being at the bottom. After considerable time, the gases in both bottles will explode at the touch of a lighted taper. Neither oxygen nor hydrogen is explosive, but their mixture is. The oxygen must have risen into the upper jar, and the hydrogen must have fallen into the lower one until a mixture was formed in both. Gases diffuse in spite of gravitation ; nor will other mechanical forces prevent it: a gas will spread in direct opposition to a current of air.

On examination it is found that hydrogen will diffuse 4 times as fast as oxygen. Now the densities of these gases are as 1 : 16 ; but notice, their diffusive rates are as 4 : 1. In other words, *their diffusive powers are inversely as the square roots of their densities.* This law applies to the diffusion of all gases.

4. *The osmose of gases.*—Porous substances can not stop the mixing of gases. Hydrogen can not for any length of time be kept pure in india-rubber bags: it will pass through the pores to mix with the air outside. The osmose of hydrogen and air may be shown very beautifully with the apparatus seen in Fig. 16. A porous cup (of Grove's battery) is fitted air-tight to a tube which reaches into colored liquid, contained in a

bottle. If a jar of hydrogen is held over the cup, bubbles instantly rush out of the tube showing that hydrogen has entered to mix with the air inside. Nor is this all: if the jar is removed, the liquid quickly rises in the tube, showing that the hydrogen has gone to mix with the air outside the cup, and that it has gone so much faster than air could flow in, that a partial vacuum has been formed.

Fig. 16.

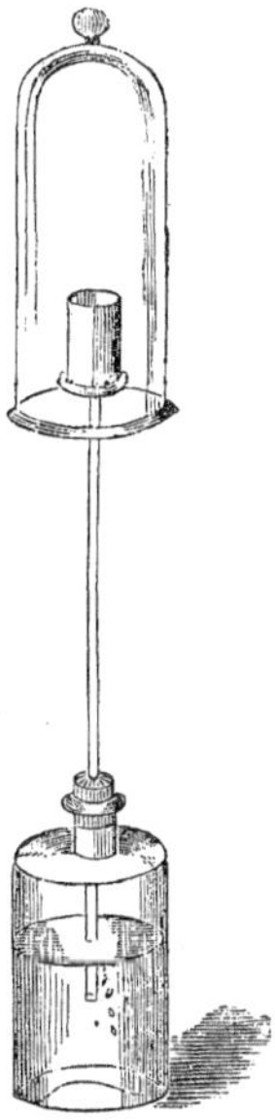

We can now see why gases of such different weight as those which form the atmosphere are kept uniformly distributed instead of forming layers, the heaviest at the bottom. Think, moreover, how vast the quantities of unwholesome, and even poisonous gases, which are given off by decaying substances, and from sewers, swamps, and marshes. Were gases subject only to the laws of gravitation, the heavy and poisonous carbonic di-oxide, with miasms and effluvia, would render animal life impossible.

(9.) Substances in a mixture may be separated by various processes; we will notice filtration, evaporation, and distillation.

1. *Filtration.*—When a solid in fine division is mixed with a liquid, it may be separated by passing the liquid through some porous substance which will not let the particles of the solid pass. The process is called filtration. The porous body through which the fluid passes is called a filter, and the clear liquid which issues is called the filtrate. A filter of common form is made

of unsized paper. Cut in the shape of a circle, it is folded so as to fit into a funnel (Fig. 17). The turbid fluid, poured upon it, filters through to be caught in the vessel below, while the sediment is left upon the filter.

Fig. 17.

2. *Evaporation.*—When a solution is gently heated the solvent slowly passes off as vapor, while the soluble solid is left behind. Exposed to the air liquids slowly evaporate and leave the solids which may have been dissolved in them. By this process only the soluble substances will be retained, the solvent passing away into the air.

3. *Distillation.*—Distillation consists in boiling a liquid and afterward condensing the vapor in a separate vessel. It may be resorted to when the solvent is a valuable substance, or when liquids are to be separated from each other. Of the first case we have an example in the common process of distilling water. The impure water is boiled and the steam is carried over from the boiler to be condensed in a receiver kept cold by a stream of water running upon it. This condensed steam is nearly pure water. Greater purity may be secured by repeating the process.

Different liquids boil at different temperatures. Water, for example, at 100° C. (212° F.) and alcohol at 80° C. Now suppose a mixture of these liquids is heated to 80° C. The alcohol will be boiled, while but small quantities of water will evaporate. The alcohol will be driven over and may be condensed in another vessel, while the water will, for the most part, remain. In this way liquids may be separated from each other.

CHAPTER II.

ON CHEMICAL ATTRACTION.

General Statement.—Attraction acting upon the atoms of bodies, under certain conditions, and subject to definite laws, produces all compound substances.

(10.) Chemical attraction can act only between unlike kinds of matter. It must be stronger than the molecular forces which oppose it in order to cause combination.

If substances in solution contain the constituents of an insoluble compound, this compound will be formed. Or if they contain the constituents of a compound which is volatile at the temperature of the solution this compound will be formed.

1. *Chemical attraction.*—By chemical attraction we mean the attraction which causes constituents to combine, and which afterward holds them in combination. Thus, when hydrogen burns in air, water is formed; it is the chemical attraction which unites the hydrogen and oxygen, and the same influence afterward holds them together in the form of water.

2. *It acts only between unlike kinds of matter.*—Two portions of hydrogen may be put together, or two portions of oxygen, but in neither case can there be a change of properties. In neither case will the action of

chemical attraction be possible even though heat or electricity be applied. But if these two kinds of gas are mixed, and then heated by a match, combination instantly follows, and water is produced by the action of chemical attraction.

Adhesion also acts only between different kinds of matter, but unlike adhesion, chemical attraction always produces new and different substances out of those on which it acts.

3. *It must be stronger than other molecular forces.*— In the solid forms of matter, cohesion is generally too strong to be overcome by chemical attraction; so too in gases, molecular repulsion is often too strong. The liquid condition, in which these forces are nearly balanced, we find to be most favorable to chemical action.

To weaken cohesive attraction is to facilitate chemical action. Potassic chlorate and sulphur, in very small quantities (to avoid danger), when rubbed together in a mortar produce a series of violent explosions. Now in this case the cohesion of these solids is overcome, while they are at the same time pressed into close contact; by this means the chemical attraction is enabled to form new compounds of their elements.

To pulverize a solid is not, however, always enough. Thus if sodic carbonate and tartaric acid, both in the finest powder, be mixed most thoroughly, or rubbed together violently, no chemical action will take place. But when a little water is added to this mixture, a violent chemical action will quickly follow. Now in this case, the water by dissolving the solids has overcome cohesion to such an extent that chemical attraction can bring about a chemical action. In this way solution very generally facilitates chemical action.

Cohesion may also be overcome by heat; for this reason heat is often applied to bring about a chemical action. Heat may also be used to cause a combination of gases among whose molecules there is a strong repulsion instead of attraction. Such is the case with a mixture of oxygen and hydrogen which explodes at the touch of a match. It may be supposed that the heat, in such cases, by increasing the vibrations of the molecules, throws those of different kinds together, and in this way, assists the chemical attraction.

4. *Bodies in solid and gaseous form may combine.*— Yet we must not suppose that no two solids can combine without their cohesion being first overcome by the methods just described. If, for example, upon a thin slice of phosphorus a crystal of iodine is laid, the two will shortly burst into a rapid and curious combustion. Nor is it true that no two gases will combine when simply brought in contact. The nitric oxide (see p. 55) when it meets with oxygen in the air instantly seizes it, and the cherry-red fumes announce the combination of these colorless gases. But such examples are quite rare; the liquid form is most favorable to chemical action.

5. *Bodies in solution do not always act chemically.*— On the other hand substances in solution do not always act chemically. A chemical action will take place, however, if they are capable of forming a new substance which is insoluble in the liquid. For example: it has been seen (p. 32) that when hydrochloric acid is added to a solution of a compound containing silver, a white precipitate is formed. Now the acid contains chlorine, but the compound of chlorine and silver is not soluble in water, and a chemical action takes place by which this insoluble compound is produced in the form of a

white precipitate. If then, one can know the constituents of the bodies in solution, and the solubility of the new compounds which they are capable of forming, he can predict whether any chemical action will take place, and if there should, what new compound would be formed.

Again; if bodies in solution have the constituents of a substance, which would be a gas at the temperature of the experiment, a chemical action will take place, and this gas will be formed. In sodic carbonate, for example, there are carbon and oxygen—the constituents of carbonic di-oxide, which we know to be a gas at common temperature. Now when tartaric acid is added to a solution of sodic carbonate a violent action follows and carbonic di-oxide is set free. Had there not been the constituents of a gas in these solutions, this chemical action would not have occurred.

(11.) Chemical attraction is governed by laws, generally called laws of combination. They may be stated in reference either to the volumes of the constituents which unite, or to their weight.

I.—COMBINATION BY VOLUME.

A.—The first law states that a compound is always formed of the same constituents in the same proportions by volume. This law may be illustrated by the composition of water, of hydrochloric acid, and of ammonia. Combining volumes are the smallest relative proportions by volume in which substances combine.

1. *The law illustrated by water.*—By the analysis of water (p. 30) the volume of hydrogen was found to be

just twice as great as that of oxygen. Now from whatever source water is taken, it is found to be made up of just these two gases, and in these same proportions—two of hydrogen to one of oxygen.

2. *The law illustrated by hydrochloric acid.*—The hydrochloric acid found in commerce is a liquid, but a simple experiment will show that this liquid is the solution of a gas in water. Some of the liquid is put into a flask (Fig. 18), and heated: a colorless gas is by this means driven over through the bent tube, and, be

Fig. 18.

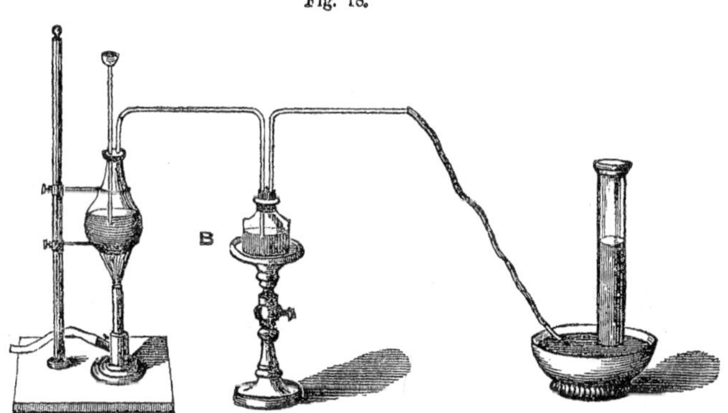

ing dried while going through sulphuric acid in the bottle B, finally enters a jar, previously filled with mercury, and inverted over a small cistern of the same fluid. If, when the jar is full of gas, it be taken from the mercury and its open mouth inserted in water, the gas will be dissolved, the water rising into the jar at the same time with surprising swiftness. Now, the solution thus obtained is found to be weak hydrochloric acid, and we hence learn that the real acid is a gas.

Of what is this acid composed? another experiment

will teach us. Into a V-shaped tube (Fig. 19) put enough of the liquid acid to fill one arm, *a*, which is closed, and partly fill the other, *b*, which is left open. There is a platinum strip in the liquid of each arm: to that in the closed end fix the negative wire of a battery and the positive wire to the other. This done, a colorless gas is seen to collect in the closed arm, and this gas, when tested, is found to be hydrogen. Now change the battery wires, fixing the positive pole to the closed arm; the hydrogen bubbles will escape into the air at *b*, and after some time a greenish gas will be seen collecting in the closed end of the tube. This gas is chlorine, which will, in due time, be carefully examined. Since no oxygen is given off in this experiment, it follows that it is the acid and not the water, which has given these gases; hence hydrochloric acid is composed of hydrogen and chlorine.

Fig. 19.

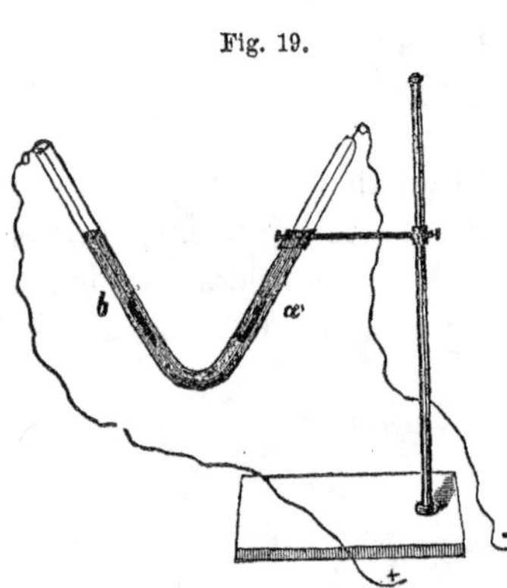

We must next find what proportion of these gases combine to form the acid. This may be done by synthesis. For this purpose, a strong, graduated glass tube, closed at one end, is used. Having been first filled with mercury and then inverted over a cistern of the same liquid, a measured quantity of chlorine and afterward another larger quantity of hydrogen is passed into it. The tube (Fig. 20) is then left several hours exposed to diffuse light, and afterward for a few moments to direct sunlight. This done, the open end is tightly shut with the finger, and the tube removed to a vessel of water under which it is again opened. The water rises quickly, until it fills a space *just twice as great* as was at first filled with *chlorine*. The remaining gas, tested with a burning taper, is found to be hydrogen, and the water contains hydrochloric acid. This shows that the two gases combine in equal volumes to form the acid.

Fig. 20.

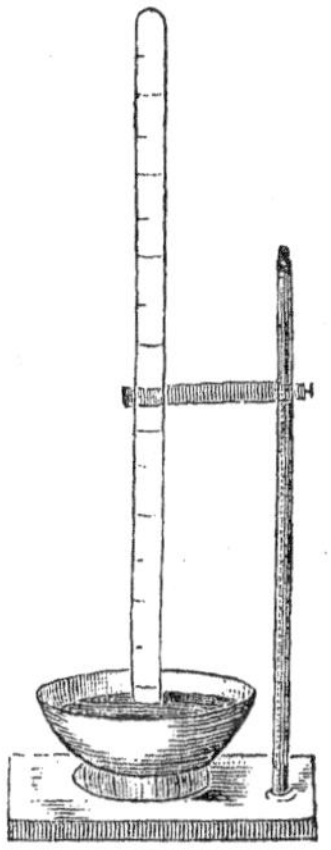

Now, by whatever means the composition of this acid is found, the same result is reached. Hydrochloric acid is always made up of hydrogen and chlorine in equal proportions by volume.

3. *The law illustrated by ammonia.*—Ammonia is made in large quantities for use in the arts, mainly from the ammoniacal liquors of gas works. It was formerly made by heating bones or other animal matter in close vessels. The horns of the deer having been largely employed for the purpose, the common name of ammonia was hartshorn. Commercial ammonia is a liquid, but the real substance is a gas, and this liquid is its solution in water. The gas is colorless, and has the well-known pungent odor of hartshorn.

By using the same apparatus (Fig. 19) and treating ammonia in just the same way that hydrochloric acid was analyzed, we learn that ammonia gas is a compound of hydrogen and nitrogen.

To determine the proportions of these two elements in ammonia, the glass tube T (Fig. 21) is first filled with pure chlorine gas. Then, by means of a funnel, *t*, hav ing a stop-cock in its neck, and fitted to the end of the tube by an air-tight joint, ammonia, drop by drop, is allowed to fall through the chlorine. After considerable liquid has collected at the bottom, the ammonia is taken from the funnel, and a little dilute sulphuric acid is put in its place, to remove the excess of ammonia in the tube. Water is then added. After no more water will enter, the tube is found just *two-thirds* full. The colorless gas, which fills the remaining one-third, is pure nitrogen, while the presence of hydrochloric acid may be detected in the fluid. Observe now: the chlorine has decomposed the ammonia and united with its hydrogen, to form the hydrochloric acid, while its nitrogen is left in the tube free. The tube full of chlorine must have taken a tube full of hydrogen, but it has set free only one-third of a tube full of nitrogen. Hence ammonia is made of three volumes of hydrogen and one volume of nitrogen.

Fig. 21.

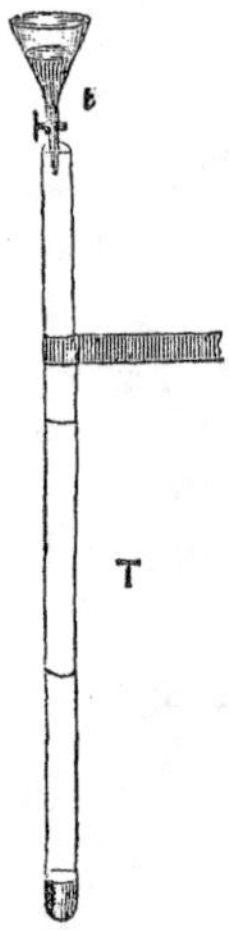

Ammonia from whatever source, or by whatever method it may be examined, *always* gives by analysis three times as much hydrogen as nitrogen, by volume.

Should we go on analyzing various compounds, we should find that each one invariably contains the same constituents, in the same proportions by volume.

4. *Combining volumes.*—Now observe the proportional volumes of the constituents in the three compounds just described:—

In Hydrochloric acid,	1 vol. of Chlorine	to	1 vol. of	Hydrogen.
" Water	1 " Oxygen	"	2 " "	Hydrogen.
" Ammonia	1 " Nitrogen	"	3 " "	Hydrogen.

The volume of hydrogen in the first is $\frac{1}{3}$ of that in the third, and $\frac{1}{2}$ of that in the second: so that if one cubic inch of hydrogen unite with one of chlorine, it will take two cubic inches to unite with one of oxygen, and three cubic inches to unite with one of nitrogen. In no known compound is the proportionate volume of hydrogen less than in hydrochloric acid. This smallest proportional volume in which hydrogen combines with other elements is called its *combining volume.* Now when the smallest volumes of other substances which enter into combination, are compared to that of hydrogen as the unit, their relative values are called their combining volumes. Thus the combining volume of ammonia, as we shall see, is 2, because the smallest volume of ammonia which can enter into combination is twice as great as the combining volume of hydrogen.

The combining volumes of the elements as far as they have been found are given in the table on p. 21.

B. The second law of combination by volume states that if one substance combines with another in more proportions by volume than one, these proportions will all be multiples of its combining volume. The compounds of oxygen and nitrogen—five in number, illustrate this law.

1. *Nitrous oxide.*—Nitrous oxide may be obtained by heating ammonic-nitrate. The nitrate is put into a flask (Fig. 22), from which a bent tube reaches over into a small bottle standing in a vessel of cold water. Another tube passes from this bottle over to a jar on the shelf of the cistern. By heat the nitrate is melted and afterward decomposed. Water and nitrous oxide

are formed. The water is condensed in the cold bottle while the oxide is collected in the jar.

Fig. 22.

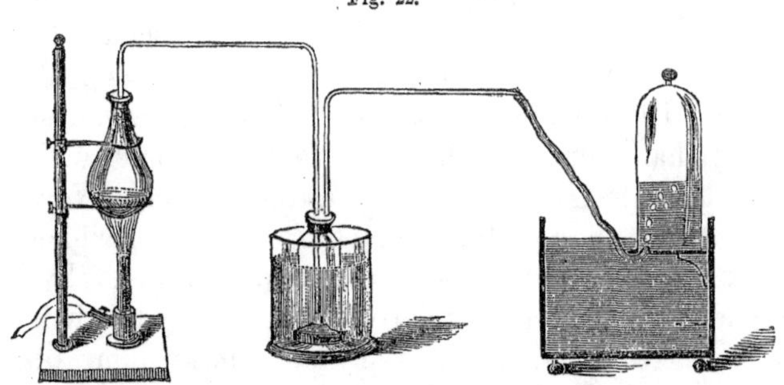

Nitrous oxide is a colorless gas, a little heavier than air. The chemical force between its constituents is weak; a lighted taper decomposes it, and taking its oxygen, burns with almost as great brilliancy as in oxygen. When breathed, its effects upon the system are peculiar. It often causes a lively intoxication, with a disposition to laughter: for this reason it has been called laughing gas. It often produces entire insensibility, and is administered for this purpose, by surgeons, to patients upon whom they are to operate. If impure, or carelessly given, it may produce death.

What is the composition of this gas? It may be determined by means of a eudiometer, shown in figure 23. Four equal divisions are marked off from the closed end of the tube. Two of these divisions are filled with nitrous oxide; the remaining two are afterward filled with pure hydrogen. By an electric spark a violent explosion is made: steam is condensed on the side of the tube, and water from the cistern will rise, leaving the two upper divisions only filled with gas: this gas, when tested, is found to be nitrogen.

Fig. 23.

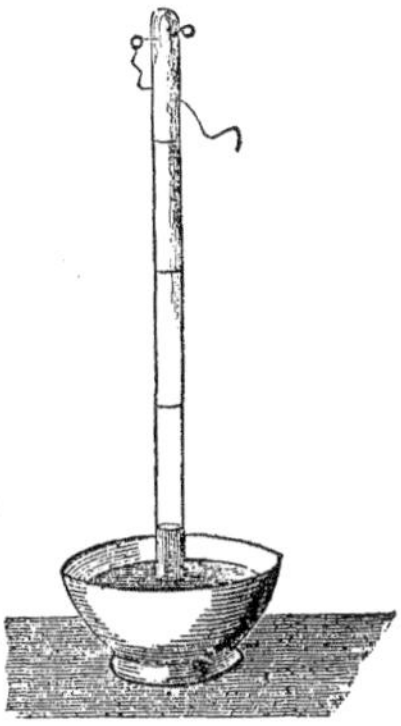

It is clear that the two volumes of hydrogen have taken oxygen enough —one volume—from the oxide, to form the water that was condensed on the tube, and have left two volumes of nitrogen. The nitrous oxide, then, was composed of two volumes of nitrogen and one of oxygen.

If we represent equal volumes of the two gases by equal squares, and their names by their symbols, the composition of the compound may be shown to the eye, by the following diagram:—

N	N	+	O

2. *Nitric oxide.*—Nitric oxide may be obtained by the action of copper upon dilute nitric acid, in an apparatus similar to that used in the preparation of hydrogen. (Fig. 24.)

The copper decomposes the nitric acid; red fumes fill the bottle, but when the nitric oxide bubbles through the water into the jar it is seen to be colorless and transparent. A lighted taper will be instantly extinguished by this gas, but burning phosphorus will decompose it, take the oxygen from it, and burn with exceeding brilliancy.

This gas is decomposed also by heated potassium, and when a measured quantity is used the composition of

Fig. 24.

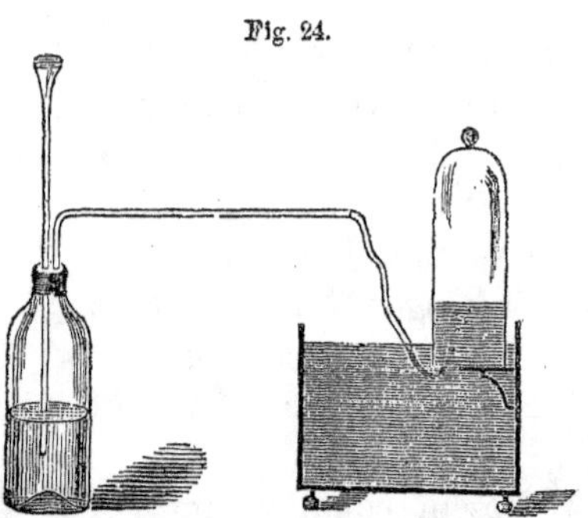

the gas may be found. It is composed of one volume of nitrogen and one volume of oxygen. It is represented to the eye by the following diagram:—

N	+	O

3. *Nitrous anhydride.*—Nitrous anhydride (nitrous acid) is a third compound of nitrogen and oxygen obtained with difficulty and imperfectly known. It has been found to consist of two volumes of nitrogen and three of oxygen. Thus:—

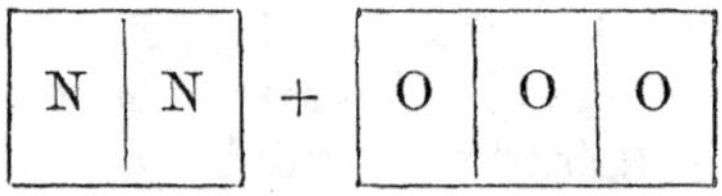

4. *Nitric peroxide.*—If nitric oxide is allowed to escape into the air, the dark cherry-red vapors which ap-

pear announce its combination with oxygen. This red substance is nitric peroxide (hyponitric acid). By measuring the volumes of nitric oxide and pure oxygen needed to produce this compound, its composition has been found to be, one volume of nitrogen to two volumes of oxygen. Hence the diagram:—

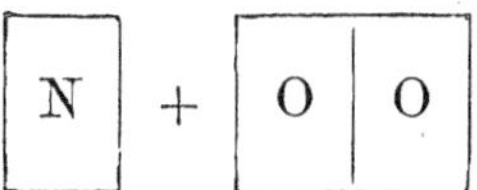

5. *Nitric anhydride.*—This substance is generally called nitric acid; it is however a very different substance from the real acid. The real acid is a compound of nitrogen, hydrogen, and oxygen; the anhydride contains no hydrogen. When analyzed it is found to be composed of two volumes of nitrogen to five volumes of oxygen. Thus:—

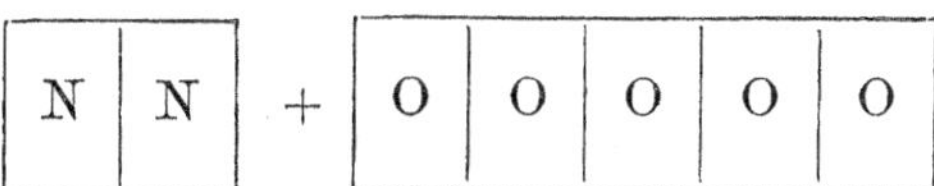

The commercial acid is a compound formed of the constituents of this anhydride and water. In combination with potash it forms niter or saltpeter, which is found often in large quantities in caves, and in small quantities scattered through the soil almost every where. From this substance nitric acid is obtained by the action of sulphuric acid.

It is a colorless and very corrosive liquid, while the anhydride is a white solid substance.

The anhydride is of no practical importance, while

the acid is one of the most useful substances of which chemistry treats. It stains the skin and other organic bodies yellow, and is used in dyeing. The yellow patterns on table-spreads are sometimes due to its action. The metals decompose it readily, and take a part of its oxygen to themselves. On this account it is used to etch copper, and is of great value in testing for the metals.

6. *The law illustrated.*—If now we examine the composition of the five compounds just described, which may be best done by writing their diagrams so that the plus signs shall be in a vertical column, we may notice that nitrogen and oxygen combine with each other in more proportions than one, the quantity of each being exactly one, two, three, or five times the quantity found in that which contains its smallest volume: there are no fractional volumes.

C.—The third law of combination by volume states that the combining volume of a gaseous or volatile compound is 2.

1. *The law illustrated.*—We have learned that hydrochloric acid is composed of one volume of hydrogen to one volume of chlorine: we have now to find the volume of the compound produced.

Into a glass tube inverted over mercury put equal volumes of the two gases and allow the apparatus to stand in diffuse light. After some hours, the greenish color of the mixture will have entirely disappeared, the gases having combined to form the colorless hydrochloric acid. The mercury stands at the same height in the tube as at the beginning of the experiment. The volume of the hydrochloric acid is therefore just equal to the volume of both constituents.

Let the following diagram represent this combination:—

H	+	Cl	=	H Cl

Experiments quite as decisive have been made to show that in water, the 2 volumes of hydrogen and 1 volume of oxygen produce only 2 volumes of water-vapor. Thus:—

H	H	+	O	=	H_2 O

And even in the case of ammonia, in which there are 3 volumes of hydrogen to 1 of nitrogen, it has been proved that there are only 2 volumes of the compound.

H	H	H	+	N	=	H_3 N

Among the gaseous compounds of nitrogen and oxygen the same thing is true: whatever the number of volumes of these gases which enter into combination, only 2 volumes of the compound will be made. So general is this result that it has come to be an accepted truth in chemistry that 2 is the combining volume of a compound. The apparent exceptions are among substances which are volatile only at a high temperature, and may be explained by supposing that these substances are decomposed by the intense heat needed to vaporize them, new compound gases being formed which again unite when the heat is withdrawn. This decomposition at high temperatures, to re-combine on cooling, is called *dissociation.*

II.--COMBINATION BY WEIGHT.

A. The first law of combination by weight states that the same compound is always formed of the same constituents in definite and invariable proportions by weight. The smallest relative proportions by weight in which substances combine are called combining weights.

1. *The law.*—By weighing the constituents obtained by analysis of different specimens of the same substance it is found that their weights in every case have the same ratio to each other. In water, for example, there will invariably be found just 8 times as much oxygen as hydrogen. Or if water is formed by synthesis, the hydrogen which enters into combination will invariably take just 8 times its own weight of oxygen. If by any means hydrogen is made to combine with more than 8 times its weight of oxygen, it will form a substance quite unlike water. Water is always composed of hydrogen and oxygen in the ratio, by weight, of 1 : 8. And so with every compound: the ratio of its constituents by weight is definite and invariable.

These proportions by weight may be calculated from the proportions by volume, if the specific gravities of the gases are known, and since the weighing of gases is difficult, this method is valuable. To illustrate: we know that a cubic inch of air at 32° F. weighs .325 gr., and that by multiplying this by the specific gravity of any gas we find the weight of one cubic inch of it. Examine hydrochloric acid. It is made of equal volumes of hydrogen and chlorine, the specific gravity of the first being .0692, of the second 2.46. Then :—

.325 gr. × .0692 = .0225, weight of 1 cub. in. of hydrogen.
.325 gr. × 2.46 = .7995, " " " chlorine.
But .0225 : .7995 : : 1 : 35.5.

Hence hydrochloric acid is composed of its constituents by weight in the ratio of 1 : 35.5.

2. *Combining weights.*—In hydrochloric acid, we may remember, the relative volumes of hydrogen and chlorine are as small as in any known compound, and we now notice that if we call the weight of one volume of hydrogen 1, that of one volume of chlorine must be 35.5. These are the *smallest relative proportions*, by *weight*, in which these two elements combine together or with others; they are called *combining weights.* The combining weights of all substances are compared with that of hydrogen, which, being the smallest, is called 1. The combining weight of chlorine is 35.5, by which we mean simply, that the smallest weight of chlorine which can enter into combination is 35.5 times greater than the smallest weight of hydrogen which can combine with other substances.

For another illustration of this important subject, let us examine the case of oxygen. The proportion of oxygen in water is as small as in any known substance. If we can find out how many times greater it is than the combining weight of hydrogen, this will be the combining weight of oxygen. In water there are 2 parts of hydrogen, and we have also learned that the oxygen is just 8 times as heavy. Being 8 times as heavy as 2 parts, it must be 16 times as heavy as 1 part: hence the combining weight of oxygen is 16.

The combining weights of the elements are given in the table on p. 21.

3. *Combining weights of gaseous elements are the weights of equal volumes.*—We have just seen that the weights of equal volumes of hydrogen and chlorine are to each other as 1 : 35.5, and that the weight of an equal volume of oxygen is 16. It will be noticed that these weights of equal volumes are the combining weights of the elements. This is very generally true of gaseous and volatile elements.

4. *Hence they represent specific gravities.*—By the term specific gravity we simply mean the relative weights of equal volumes of different substances. Air is very commonly the standard with which to compare gases, and the specific gravity of a gas tells how many times heavier it is than an equal bulk of air. But among chemists *hydrogen is the standard*, and the specific gravity of a gas tells how many times heavier it is than an equal volume of hydrogen.

Now, calling the weights of a given bulk of hydrogen 1, the weights of equal bulks of other elementary gases are shown by their combining weights. Hence the number that represents the combining weight of an element represents its specific gravity also.

There are exceptions to this. The combining weight of phosphorus is the weight of one-half a volume of phosphorus vapor: the same is true of arsenic. In these cases the specific gravity is twice the combining weight. On the other hand, the combining weight of mercury is the weight of a double volume: the same is true of cadmium. In these cases the specific gravity is one-half the combining weight.

Of most of the elements, solid at ordinary tempera

ture, the specific gravity of their vapors has not been determined.

5. *The specific gravity of a compound gas is one-half its combining weight.*—We have seen that the combining volume of a compound gas is 2. The combining weight of it is thus the weight of two volumes. But specific gravity is always the weight of one volume, and hence, of a compound gas, it must be one-half the combining weight.

The combining weight of hydrochloric acid is 36.5; its specific gravity is $\frac{36.5}{2} = 18.25$. The combining weight of ammonia is 17; its specific gravity is $\frac{17}{2} = 8.5$. Air is 14.4 times heavier than hydrogen. By dividing the specific gravity of any gas by 14.4, we get its specific gravity compared with air; or, multiplying its specific gravity on the air standard by 14.4, will give its specific gravity on the hydrogen scale.

B. The second law of combination by weight states, that if one substance combines with another in more than one proportion by weight, these proportions will always be multiples of its combining weight.

1. *The law illustrated.*—We have seen that oxygen and nitrogen form five different compounds. Their composition by weight has been found by analysis to be as follows:—

Nitrous oxide,	28	of nitrogen	to	16	of oxygen,
Nitric oxide,	14	"	"	16	"
Nitrous anhydride,	28	"	"	48	"
Nitric peroxide,	14	"	"	32	"
Nitric anhydride,	28	"	"	80	"

Now, 14 is the combining weight of nitrogen, and 16 is that of oxygen; and we notice that the proportions of nitrogen are all multiples of 14, those of oxygen all multiples of 16. And so it will ever be found; there can be no fractional parts of the combining weight of a substance.

C. The third law of combination by weight states, that the combining weight of a compound is the sum of the combining proportions of its constituents.

1. *The law illustrated.*—Water combines with other substances, and the smallest proportion is invariably just 18 times as great as the combining weight of hydrogen; hence its combining weight is 18. But it consists of two combining weights of hydrogen and one of oxygen. The sum of these, $2+16$, is 18. Hence the combining weight of water is the sum of the combining proportions of its constituents.

It is important to notice that the term combining proportions, as just used, does not mean combining weight in all cases. The combining weight of hydrogen is 1: the quantity which combines with 16 of oxygen, however, is 2, and it is this which enters into the combining weight of water. We have applied the term combining weight to the *smallest* proportion of any substance which may enter into combination: we shall apply the term combining proportion to the relative weight actually existing in the compound. Thus the combining weight of nitrogen is 14: its combining proportion in nitrous oxide is 28.

Again: Nitric anhydride combines with other substances; what is its combining weight? It consists of two combining weights of nitrogen and five of oxygen. The combining proportions are 2×14 or 28 of nitrogen and 5×16 or 80 of oxygen; the sum of these, $28+80=108$, is the combining weight of the compound.

(12.) The laws of combination are independent of all theories, having been established by repeated and decisive experiments. The "atomic theory," however, has been proposed to explain them, and because it does explain them better than any other, it is generally accepted.

I.—THE THEORY.

A. The atomic theory assumes:—

1. That all bodies are made up of molecules, and that these, in turn, consist of indivisible atoms.
2. That all atoms of the same kind have equal weight.
3. That combining weights are the relative weights of the atoms of different substances.
4. That compounds are formed by the union of different kinds of atoms.
5. That the nature of a compound depends upon the kind, number, and arrangement of its atoms.

1. *The molecule.* Professor Hoffman* has given the most clear and elegant definition of the present views of chemists in regard to the composition of matter. We can not quote in full; among other things, he says: "However finely we may grind up ice, for example; if we took care to keep the temperature below the freezing point we should still have blocks of ice. Our finest ice-powder would still consist of very small frag-

* See Hoffman's "Introduction to Modern Chemistry," or, The Chemical News.—Am. Rep., vol. i, p. 217

ments of solid ice; and if, of this ice-dust, we took the smallest grain, we could, by applying heat, turn it into water, thus proving it to have *parts* capable of separation; and further, the smallest possible portion of this water, by being heated, is expanded into steam, showing that it was likewise made of still smaller parts." Now in these changes from ice to water, and from water to steam, we have produced no change in the nature of the substance. The little particles of the steam, existed at first in the block of ice. The steam particles, however, can not be divided *without changing their nature.* "They are *the smallest portion of this kind of matter which can exist in a free state.*" They are called *molecules.* It is believed that all bodies are made up of molecules, separate bodies, but so small as to be far beyond the reach of the most powerful microscope.

2. *Atoms.*—But these molecules are not indivisible. Steam, if passed through a red-hot iron tube is decomposed into the gases hydrogen and oxygen. This must be as true of one molecule of steam as of any other quantity, hence the molecule of water is made up of still smaller parts of the gases named. "Here the divisibility of matter, so far as our experimental knowledge goes, reaches its final term. The elements are, as we remember, so called, precisely because they resist every agency which we can bring to bear in the hope of decomposing them." The smallest portions into which we can conceive the elementary bodies to be divided are called *atoms.*

3. *Atoms of the same kind are alike.*—It is thought that atoms of the same kind of matter—of oxygen, for example—are in all respects, size, shape, and weight, alike. Atoms of different kinds, however, have differ-

ent weights. One of oxygen is supposed to be 16 times heavier than one of hydrogen, and, in general the combining weights of the elements are the relative weights of their atoms. Hence the term "atomic weight" is often used instead of combining weight.

4. *Molecules of compound gases all the same size.*—We have learned in natural philosophy that all gases expand alike by equal additions of heat, and contract alike when cooled; moreover, that the volumes of all alike are inversely as the pressure upon them. Now any change in volume of a gas must be due to a change in the distance between the molecules,—in expansion they vibrate through greater distances, in contraction they vibrate through less distances, and hence are brought nearer together. And since all gases are affected exactly alike by the forces of heat and pressure, it is inferred that their molecules vibrate through equal distances; or, in other words, the distances between their molecules are alike. This idea is expressed in the following law:—

Equal volumes of all gases, at the same temperature and pressure, contain the same number of molecules.

If this is true, then the molecules of all true gases must be of the same size.

5. *Simple as well as compound gases.*—We have said that *all* gases are affected by heat and pressure alike. The element hydrogen, and the compound hydrochloric acid are expanded and contracted in the same way; neither does chlorine differ sensibly from either. The inference is that any volume—say one cubic inch—of hydrochloric gas contains just as many molecules as the same volume of hydrogen or of chlorine.

Now let us trace this thought to its conclusion. We

have seen that hydrochloric acid consists of hydrogen and chlorine. Every *molecule* of the acid contains *one atom* of hydrogen and one of chlorine. We have also seen that one cubic inch of hydrogen with an equal volume of chlorine forms two cubic inches of the acid. There must then be as many *atoms* in one cubic inch of hydrogen as there are of *molecules* in two cubic inches of hydrochloric acid. In other words; there are twice as many atoms of hydrogen in a cubic inch as there are molecules in the same volume.

But, according to the law, the number of molecules in equal volumes of the two gases is the same. Hence there are twice as many *atoms* of hydrogen in a cubic inch as there are *molecules* of hydrogen in the same volume. It must therefore take *two atoms* to make *one molecule* of hydrogen.

And so the chemist comes to believe that even the elementary gases, in a free condition, are made up of molecules—each molecule being a group of at least two atoms. All matter, then, whether simple or compound, consists of molecules, in the element the atoms of the group are all alike, in the compound they are of different kinds.

When elements combine, their molecules are broken up; the atoms of one combine with the atoms of another to form a compound molecule.

A molecule, then, is the smallest portion of any kind of matter that can exist in a free state; an atom is the smallest portion of an element which we can conceive of even in combination.

II.—APPLICATION OF THE THEORY.

B. The atomic theory furnishes an explanation of the laws of chemical combination and of the phenomena of isomerism and allotropism.

1. *Of the first law of combination.*—According to the theory a compound is formed by the union of atoms, and its nature depends partly upon their number; and at the same time each one has a definite weight. Hence a compound is formed of definite and invariable weights of its constituents.

2. *Of the second law of combination.*—According to the theory, elements can unite only by atoms, and the atoms are not divisible. More proportions of an element than one is possible in combination, only because a different number of whole atoms may combine with those of another element. But the weight of any number of whole atoms must be a multiple of the weight of one. Hence, if one element unite with another in more proportions than one, these proportions will all be multiples of the combining weight.

3. *Of the third law of combination.*—The molecule of a compound is made up of atoms of its elements: these have a definite and unchangeable weight; hence, the weight of the molecule must be the sum of the weights of its atoms. But the molecule of the compound is the smallest portion of it that can enter into combination; hence, the combining weight of a compound is the sum of the combining proportions of its elements.

4. *Isomerism.*—It is a curious fact that the same elements and in the same relative proportions, do not always form the same compound. For example, starch,

which is insoluble in water, and dextrine or British gum, very soluble, the two differing also in other properties, are both composed of carbon, hydrogen, and oxygen—6 combining weights of carbon, 10 of hydrogen, and 5 of oxygen. Substances having the same composition, but different properties, are said to be *isomeric.*

The only explanation that can be given of this curious phenomenon is found in the assumption that the nature of a compound depends upon the arrangement of its atoms as well as upon their kind and number.

5. *Allotropism.*—We have seen that an element may exist in different conditions with different properties. Oxygen and ozone are but different forms of the same element. This property possessed by some elements is called *allotropism.* The diamond, plumbago, and charcoal, with all their striking differences, are still but allotropic forms of the element carbon.

The only explanation of this curious property, seems to be the assumption that it is due to a different method of grouping the atoms in the molecules or of grouping the molecules among themselves.

The term allotropism was formerly applied to elements only, but it has come to be applied to compounds also. In all cases of isomerism the differences between compounds are so great as to lead chemists to give a distinct name: but where the differences are too slight to warrant this, the two substances are called allotropic forms.

Allotropism is often associated with different crystalline forms. The crystals formed by allowing melted sulphur to cool slowly are in the form of long slender needles, very different indeed from the form in which the crystals (rhombic octohedra) of this substance are found in nature. In a third condition, obtained by heating sulphur to about 230°C., and then pouring it into cold water, it shows no crystalline form whatever. These three varieties are called allotropic forms of sulphur.

(13.) The effect of chemical attraction is to produce new substances, either by causing direct combination or by substitution. This effect is announced by a change in color, temperature, form or other properties.

1. *By direct combination.*—Direct combination takes place when a new compound is formed without any previous decomposition. The union of hydrogen and oxygen, when their mixture is touched with a burning match, is a familiar example. The elements in the mixture are free, and when the proper temperature is reached, their combination is the only chemical action.

Not only elements, compounds also may enter into direct combination. The slaking of lime is a familiar case. Lime has a very strong attraction for water. When the two substances are brought in contact a chemical action occurs, announced by the swelling of the lime, its crumbling to powder, and the formation of clouds of steam. The two substances combine, without other chemical action, and form slaked lime.

2. *By substitution.*—Direct combination is of rare occurrence. The production of new compounds by substitution is more common. Upon the surface of water drop a piece of potassium: it instantly takes fire (Fig. 25), runs swiftly about over the surface of the

water, and at last disappears. Even though the water had been pure at first, yet after the experiment it may be found to contain potash. This new substance is the compound of potassium, oxygen, and hydrogen. The potassium has simply taken the place of one part of hydrogen in the water. We may show this to the eye by using the symbols of the elements. Thus:—

Fig. 25.

Water, represented by $\left.\begin{matrix}H\\H\end{matrix}\right\}O$; 2 of hydrogen to 1 of oxygen becomes

Potash, represented by $\left.\begin{matrix}K\\H\end{matrix}\right\}O$; 1 of potassium substituted for 1 of hydrogen.

In this case one new compound is made while the element, hydrogen, is set free. When the action is between two compounds, it generally happens that two new compounds are made. To illustrate this let a stream of sulphuretted hydrogen gas be passed through a solution of arsenic oxide: a yellow precipitate will be formed. The explanation is this:—

Sulphuretted hydrogen consists of $\left\{\begin{matrix}\text{Sulphur,}\\\text{Hydrogen.}\end{matrix}\right.$

Arsenic oxide consists of $\left\{\begin{matrix}\text{Arsenic,}\\\text{Oxygen.}\end{matrix}\right.$

In the action which takes place, the hydrogen and arsenic change places, making two new compounds:

Arsenic sulphide consisting of $\left\{\begin{matrix}\text{Sulphur,}\\\text{Arsenic.}\end{matrix}\right.$

Water consisting of { Hydrogen, Oxygen.

The first of these appears as a yellow precipitate.

Chemical changes are very generally called reactions.

3. *Indicated by change of color.*—The production of new compounds is, very often, as in the experiment just described, shown by change of color: it was the yellow appearance which announced the chemical action. Or try another experiment. Let a solution of sugar of lead and another of sulphuretted hydrogen, both of which are as colorless as pure water, be mixed in a goblet: quickly a dense black precipitate appears. The color shows that a new black compound—lead sulphide, has been made.

4. *Indicated by change of temperature.*—In the slaking of lime, already mentioned, the new compound is made by the union of lime and water. This chemical action is accompanied by a rise of temperature, enough to change a part of the water into steam.

Or let the following experiment still further illustrate the curious fact that heat is evolved by chemical action Into some water held in a beaker glass, pour about four times as much strong sulphuric acid. So strong a heat will be at once produced, that ether, and even water in a test tube placed in the mixture, may be boiled. This strong heat announces the combination of the water and the sulphuric acid.

5. *Indicated by change of form.*—Chemical action is often followed by a change in the physical form of substances. The two *gases*, hydrogen and oxygen, unite to form the *liquid*, water. The two liquids—solutions of sugar of lead and sulphuretted hydrogen—produce the solid precipitate of lead sulphide—this black pre-

cipitate, like all others, being a solid substance in a state of very fine division.

6. *Other properties.*—Beside these changes of color, temperature, and form, many others indicate the action of chemical force. In general terms, we may say that the effect of chemical force is to produce new compounds, and that this effect is indicated by changes of properties. Decomposition can hardly be regarded as the immediate effect of chemical force; in all cases of substitution it is an apparent effect, doubtless due to the stronger force between the constituents of the new compounds. In other cases decomposition occurs because the chemical force is overcome by some other, as when electricity decomposes water; the decomposition occurs because electricity in this case is stronger than chemical force.

(14.) The names of chemical compounds are not arbitrarily chosen: they are so constructed that they show the composition of the compounds to which they belong.

I.—ACIDS.

A.—Acids are of two classes, oxacids and hydracids. The names of oxacids are characterized by the terminations *ic* and *ous* followed by the word acid. The names of hydracids are known by the prefix *hydro*.

1. *Acids.*—Let us attend to the following experiment. Into a solution of blue litmus (a vegetable blue coloring matter), put a few drops of hydrochloric acid: its fine blue color is at once changed to a bright red. Any other soluble *acid* would have caused the same change, and this is the most ready means to determine whether a substance belongs to this class of bodies.

Beside this power to change vegetable blue colors to red, acids have certain other properties in common, among which we notice *first*, that they are generally sour to the taste; *second*, that they are usually composed of non-metals; and *third*, that hydrogen is one constituent.

2. *They are of two classes.*—A large number of acids contain both oxygen and hydrogen in combination with another non-metal. In a smaller number hydrogen alone is combined with the other non-metal. Those of the first class are called *oxacids;* those of the second are *hydracids.* Since hydrogen is a constituent of both classes the word acid is enough to show the presence of hydrogen in the substance to which the name is given.

3. *The names of oxacids.*—The ending of the name of the element with which the hydrogen and oxygen are combined, is changed to *ic* or *ous*, and then followed by the word *acid.* The *ic* always denotes a larger proportion of oxygen than the *ous.*

Thus an acid compound of chlorine, oxygen, and hydrogen, is chlor*ic* acid: another, which contains a smaller proportion of oxygen is called chlor*ous* acid.

Since more than two acids may be formed of the same elements the prefixes *per* and *hypo* are used—per always indicating a larger proportion of oxygen than hypo.

Thus: an acid compound of chlorine, having more oxygen than the chloric acid, is called *per*chloric acid: one having less than the chlorous acid is called *hypo*-chlorous acid.

This system of naming the oxacids may be shown to the eye by the following skeleton. The blank may be filled with the name or an abbreviation of the name of any element which combines with hydrogen and oxygen to form an acid.

Per———	ic	acid.
———	ic	"
Hypo———	ic	"
———	ous	"
Hypo———	ous	"

From the name of an oxacid we ought to be able to know its constituents; or knowing the constituents we should be able to construct its name.*

EXAMPLES.

1. What are the constituents of phosphoric acid?
Ans. Hydrogen, oxygen and phosphorus.

By what part of the name is each one of these elements suggested?

2. What are the constituents of bromic acid?
3. Name the elements in sulphurous acid.
4. Name the elements in hyposulphurous acid.
5. What difference in the composition of the last two acids named, indicated by their names?
6. What difference in composition is indicated by the names iodic acid and periodic acid?
7. What acid will be formed by the union of hydrogen, oxygen, and bromine?
8. Name the acid which contains oxygen, manganese, and hydrogen. *Ans.* Manganic acid.
9. What other acid with the same elements, but with a larger proportion of oxygen?

* Examples like the following should be multiplied by the teacher until the pupil is familiar with the principles of the nomenclature.

4. *The names of hydracids.*—But all acids do not contain oxygen. Some consist of hydrogen and a single other non-metal; such is hydrochloric acid already so familiar to us. The names of these acids also end in *ic*, but they are especially characterized by the prefix *hydro.*

The hydracids always contain one combining weight of hydrogen to one of the other element. On this account no other endings or prefixes are necessary in the name. We recognize *hydro bromic acid* as the name of an acid compound of hydrogen and bromine.

EXAMPLES.

1. What are the elements in hydrofluoric acid?
2. Name the elements in hydriodic acid.
3. Name the elements in hydrosulphuric acid.
4. What is the difference in composition of hydrosulphuric acid and sulphuric acid?
5. Name the acid compound of hydrogen and iodine.

II.—ACID ANHYDRIDES.

B.—Acid anhydrides may be described as compounds left after taking the elements of water away from oxacids. They retain the name of the corresponding acids except that the term anhydride is used in place of acid.

1. *Acid anhydrides.*—If by any means an oxacid is deprived of its hydrogen, it will at the same time give up a part of its oxygen, the two elements being given off in the proportions to form water. Now acids from which the elements of water have been taken are no longer to be called acids, because they no longer have

acid properties. The term *anhydride* has been applied to them. For example, one compound of sulphur, oxygen, and hydrogen is called sulphuric *acid*, but if all the hydrogen with enough oxygen to form water be taken away, the remaining compound of sulphur and oxygen is called sulphuric *anhydride*. In the same way phosphoric acid deprived of the elements of water becomes phosphoric anhydride.

Most of the anhydrides will combine with water and produce acids, and many of the acids when heated will give up water and become anhydrides.

EXAMPLES.

1. By what change in composition would nitric acid become nitric anhydride?
2. Name the constituents of nitric anhydride.
3. What are the elements in phosphoric anhydride? Wherein does it differ from phosphoric acid?
4. What acid would give nitrous anhydride by losing the elements of water?
5. Name the acid formed by adding the elements of water to sulphurous anhydride.

III.—BASES.

C.—Bases are called hydrates. And to show which hydrate is meant in any case, the name of its metallic constituent with its ending changed to *ic* or *ous*, is used as an adjective.

1. *Bases.*—By the following experiment we learn one characteristic of the class of bodies, called bases. Into the goblet of litmus solution, reddened by hydrochloric acid, in a former experiment, put a few drops

of ammonia; very quickly the fine blue color of litmus is restored.

Besides this power to restore the blue color of reddened litmus, bases have certain other qualities in common. We notice *first*, that they are generally caustic to the taste; *second*, that they are composed of hydrogen, oxygen, and a metal.

2. *The names of bases.*—The bases are called *hydrates*, and each hydrate is named from the metal it contains by changing the termination of its name to *ic* or *ous*, and then using it as an adjective. Thus hydrogen, oxygen, and potassium form a base, called *potassic hydrate:* the *ium* of the name of the metal is changed to *ic*, and followed by the word hydrate. So also, hydrogen, oxygen, and sodium form a base: from the name of the metal, sodium, we get *sodic*, and by adding hydrate, we have the desired name *sodic hydrate.*

The Latin name of the metal is often used in preference to its English name. Iron, for example, forms two hydrates,—*ferric* hydrate and *ferrous* hydrate, the Latin name of iron being *ferrum*. (See Cooke's Chem Phil., Part I., 35 and 49.)

EXAMPLES.

1. What are the elements of calcic hydrate?
2. What are the elements of magnesic hydrate?
3. What are the elements of cupric hydrate?
4. Name the elements of argentic hydrate.
5. Name the hydrate containing barium. What other elements does it contain?

IV.—BASIC ANHYDRIDES OR METALLIC OXIDES.

D.—Basic anhydrides may be described as compounds, left after taking the elements of water away from hydrates. In their names, the term oxide is used instead of hydrate.

1. *Basic anhydrides.*—Many hydrates, when heated, give off all their hydrogen, with oxygen enough to form water. The compounds left behind are very different from the original hydrates, and should be called by a different name. As acids, when deprived of water, are called acid anhydrides, so bases, when deprived of water, might be called basic anhydrides. They were formerly thought to be the true bases, and the name still clings to them—hydrates and anhydrides being still classed together as bases. They are in general, however, called *metallic oxides.*

These metallic oxides are distinguished from each other by the name of the metal, in each case changed to an adjective, as in the case of hydrates. Besides the endings *ic* and *ous*, to show different proportions of oxygen, prefixes are also used to show the number of combining weights: *di*, meaning 2; *tri* meaning 3; and such others as the case may demand.

The compound of potassium and oxygen, for example, is called *potassic oxide.* There are two oxides of barium: one contains one combining weight of oxygen to one of barium, the other two of oxygen to one of barium. The first is called *barous oxide*, the second *baric oxide.*

Or, using the prefixes, the oxides of manganese will illustrate, thus:—

Manganic *mon*-oxide = 1 of Mn. to 1 of O.
Manganic *di*-oxide = 1 " " 2 "
Manganic *sesqui*-oxide = 2 " " 3 "

EXAMPLES.

1. Mercury combines with oxygen; what shall we call the compound? *Ans.* Mercuric oxide.

2. But there is another oxide of this metal, containing a less *proportion* of oxygen; what shall it be named?

3. What are the elements of cupric oxide?

4. What are the elements of cuprous oxide?

5. What are the elements and their proportion in chromic tri-oxide?

V.—NEUTRAL BINARY COMPOUNDS.

E.—Neutral binary compounds are named by the same method as the metallic oxides.

1. *Neutral bodies.*—Substances which, like water, will neither redden vegetable blue colors, nor restore the blue after it has been reddened by an acid,—which, in a word, do not have the properties of either an acid or a base, are called *neutral* bodies. Water is a perfectly neutral body. Nitric oxide and nitrous oxide are neutral bodies.

2. *Binary compounds.*—Compounds of two elements only, are called *binary* compounds. Water is a binary compound, because made of two elements—hydrogen and oxygen. The hydracids are binary compounds. There are several neutral binary compounds of non-metals only, such as nitric oxide and carbonic

oxide, but of far the greater number one of the constituents is a metal.

3. *Their names.*—In compounds of metals with non-metals, the non-metal is the electro-negative constituent. In naming such binary compounds the ending of the name of the electro-negative element is changed to *ide;* and then the name of the other element with its ending changed to *ic* or *ous*, is used as an adjective. Prefixes, *di*, *tri*, *sesqui*, and others, are also used to indicate the number of combining weights.

For example, chlorine unites with the metals: it is the electro-negative element of the compounds; so the name chlor*ine* is changed to chlor*ide*, and these compounds are *all* called chlorides. To know which one is meant, the name of the other element must be used as an adjective. If the compound be of chlorine and potassium, the name is *potassic chloride.* So sulphur forms sulphides: with sodium it forms *sodic sulphide.* Ferrous sulphide is the compound of iron and sulphur; ferric sulphide is another; the last contains a larger proportion of sulphur than the first.

EXAMPLES.

1. Name the compound of potassium and iodine.
2. Name the compound of lead (Latin, *plumbum*) and sulphur. *Ans.* Plumbic sulphide.
3. Name the compound of lead and iodine.
4. Name the compound of copper and chlorine.
5. What are the elements in arsenic sulphide?
6. What are the elements in zincic sulphide?
7. What is the difference between cupric chloride and cuprous chloride?

8. Name the compound of gold (*aurum*) and chlorine, containing one combining weight of the metal to three of the chlorine. *Ans.* Auric tri-chloride.

VI.—SALTS.

F.—Salts may be described as compounds formed by substituting a metal for a part or the whole of the hydrogen in the acid.* If formed from the hydracids they are named by the method for neutral binary compounds. If formed from oxacids their names are characterized by the endings *ate* and *ite.*

1. *Salts.*—Notice the following experiment. Into a bottle put a few clippings of zinc, and upon them pour a quantity of hydrochloric acid. A vigorous boiling quickly begins; hydrogen gas escapes; the zinc slowly disappears and finally a clear and quiet liquid remains. Evaporate this liquid and a white solid will be left. The explanation is this: the zinc has decomposed the acid taking the place of the hydrogen which was in combination with the chlorine. Thus: using the *symbols* of the *elements,*

$$\left.\begin{array}{ll}\text{Hydrochloric acid,} & \text{H Cl}\\ \text{Zinc,}\quad . \quad . & \text{Zn}\end{array}\right\}\ \text{become}\ \left\{\begin{array}{ll}\text{Zn Cl,} & \text{Zincic chloride}\\ \text{H,} & \text{Hydrogen.}\end{array}\right.$$

The zincic chloride is the white solid left by evaporation, and the hydrogen went off into the air. The zinc has taken the place of the hydrogen and the *salt,* zincic chloride, is formed. If sodium takes the place of

* Salts may also be described as coming from bases, and in other ways which it is not thought best to develop in this elementary work. (See Cooke's Chemical Philosophy, Part I., pp. 83, 86, and 104.)

hydrogen in the same acid, the *salt*, sodic chloride (*common salt*), is formed.

Again: if the two combining weights of hydrogen in sulphuric acid are both replaced by sodium, a salt, sodic sulphate, will be formed; or if only one of them is replaced by sodium, the remaining compound is still a salt. From these illustrations we gather this description of a salt. It is a compound formed by substituting a metal for either the whole or a part of the hydrogen in an acid.

2. *The names of salts.*—The salts formed from hydracids are named by the method already given for neutral binary compounds. These salts were formerly called *haloid salts.*

Of salts formed from oxacids the nâme is made by changing the ending of the name of the acid from which they are derived, from ic to *ate* or from ous to *ite*. In case only a part of the hydrogen is displaced, the presence of the remainder is indicated in the name by the prefix hydro. When, for example, all the hydrogen of sulphuric acid is replaced by sodium, the salt is called *sodic sulphate;* but if only one of the two combining weights of hydrogen is displaced, the presence of the hydrogen left may be shown by the name *hydro-sodic sulphate.*

In the same way potassium and sulphuric acid may form either potassic sulphate or hydro-potassic sulphate.

EXAMPLES.

1. Name the salt from nitric acid and potassium.
2. Name the salt from copper (cuprum) and sulphuric acid.
3. Name the salt from hypochlorous acid and sodium

4. Name the salt from acetic acid and lead (plumbum).

5. Name the acid and the metal from which calcic carbonate may be derived.

6. Name the acid and the metal from which calcic hyposulphite may be derived.

7. What are the *elements* in ferrous sulphate?

8. What are the elements in baric sulphite?

9. Of what elements is magnesic carbonate composed?

10. What acid is required with copper to form cupric nitrite.

11. What are the constituents of potassic chlorate?

12. Name the metal in the alluminic silicate.

13. Name the metal and the acid in magnesic citrate.

14. Name the elements in calcic phosphate.

(15.) Some important exceptions to the foregoing rules of nomenclature need to be noticed. 1st. Certain metallic oxides are named by simply changing the termination of the name of the metal to *a*. 2d. Certain binary compounds of hydrogen have specific names end ing in *uretted*. 3d. Organic acids are named arbitrarily. These exceptions are still sanctioned by custom.

1. *Examples of 1st exception.*—Of several metallic oxides the names in common use are made by changing the ending of the name of the metal to *a*. Sodium and oxygen, for example, form *soda*—the termination *ium* of sodium being changed to *a*. The common name of potassic oxide is potassa; of magnesic oxide is magnesia; of baric oxide is baryta—the *yt* here being used for the sake of euphony. These names have long been in use, and are still sanctioned by the common use of chemists.

The oxides named in this way are those which, with the elements of water, form the most powerful bases.

2. *Examples of the 2d exception.*—The compound of hydrogen and sulphur has a slightly acid reaction, and as an acid it is, by the rule, called hydrosulphuric acid. Without regard to its acid properties it would be called *hydric sulphide.* The name in more common use, however, is *sulphuretted hydrogen.* Other compounds of hydrogen are named in the same way. Hydrogen with phosphorus forms phosphuretted hydrogen; with arsenic it forms arseniuretted hydrogen, and with antimony it forms antimoniuretted hydrogen. These, rather awkward names, are still in common use.

3. *Examples of the 3d exception.*—A long list of acid compounds may be obtained from vegetable and animal substances; they are called *organic acids.* Their constituents are hydrogen, oxygen, and carbon. Their specific names have been given without rule, except that they have the common termination *ic.* Acetic acid, tartaric acid, oxalic acid, are examples. Their very great number, and the fact that they are made of the same elements seem to forbid any attempt to indicate their composition by prefixes and terminations.

(16.) The foregoing principles of nomenclature have been lately adopted. The difference between the new and the old nomenclature may be seen by comparing the names of the same substances, according to the two systems.

1. *The nomenclature.*—Not until the year 1787 was any attempt made to reduce the language of chemistry to a system. But the number of compounds to be de-

scribed increased to such an extent that even the strongest memory could not hope to keep their meaningless names. To avoid this difficulty, a system was proposed by Lavoisier, by which the name of a substance should indicate its composition. The simplicity of the system, and the accuracy with which it expressed chemical theories, secured its universal adoption; but the theories themselves having now, in great part, been rejected, a new system of names is needed to represent the new theories which have been established in their stead.

2. *New and old names.*—The difference between the new names and the old may be best seen by carefully comparing them in the following lists. Such a comparison is all the more necessary to the student, because the old names are still in quite common use. Old names will linger long after the reason for their adoption has ceased to be in force. How well is this illustrated by such trivial names, as saltpeter (potassic nitrate), Glauber's salts (sodic sulphate), and oil of vitriol (sulphuric acid), which, though given before any general rules were established, still remain in use! And so we may expect that the names on the old system will long linger, while in all the writings of modern chemists the new system only is employed.

NEW NAMES.	OLD NAMES.
Nitrous Anhydride.	Nitrous Acid.
Nitric Anhydride.	Nitric Acid.
Nitrous Acid.	Nitrous Acid.
Nitric Acid.	Nitric Acid.
Carbonic Anhydride.	Carbonic Acid.
Sulphurous Anhydride.	Sulphurous Acid.

Sulphuric Anhydride.	Sulphuric Acid.
Bromic Anhydride.	Bromic Acid.

It will be noticed that the names of the acids are alike in the two systems, and that the old system makes no distinction between acids and anhydrides.

NEW NAMES.	OLD NAMES.
Sodic Oxide.	Oxide of Sodium.
Potassic Oxide.	Oxide of Potassium.
Calcic Oxide.	Oxide of Calcium.
Strontic Oxide.	Oxide of Strontium.
Baric Oxide.	Oxide of Barium.
Aluminic Oxide.	Oxide of Aluminum.
Magnesic Oxide.	Oxide of Magnesium.
Zincic Oxide.	Oxide of Zinc.
Cadmic Oxide.	Oxide of Cadmium.
Ferrous Oxide.	Protoxide of Iron.
Ferric Oxide.	Sesquioxide of Iron.
Nitrous Oxide.	Protoxide of Nitrogen.
Nitric Oxide.	Deutoxide of Nitrogen.
Nitric Peroxide.	Peroxide of Nitrogen.

It will be seen that in the names of oxides by the new system the term *oxide* is *preceded* by the name of the metal changed to an adjective, while by the old system it is *followed* by the name of the metal: the two being joined by the word *of*.

NEW NAMES.	OLD NAMES.
Sodic Hydrate.	Hydrate of Soda.
Potassic Hydrate.	Hydrate of Potassa.
Calcic Hydrate.	Hydrate of Lime.

Strontic Hydrate.	Hydrate of Strontia.
Baric Hydrate.	Hydrate of Baryta.
Aluminic Hydrate.	Hydrate of Alumina.
Magnesic Hydrate.	Hydrate of Magnesia.
Ferric Hydrate.	Hydrated Sesquioxide of Iron.

It will be seen that in naming the bases, the new system uses the term *hydrate*, preceded by the name of each particular metal, changed to an adjective, while in the old method they are "*hydrates of*" the oxides of the metals.

NEW NAMES.	OLD NAMES.
Sodic Chloride.	Chloride of Sodium.
Sodic Iodide.	Iodide of Sodium.
Potassic Bromide.	Bromide of Potassium.
Ferrous Sulphide.	Protosulphide of Iron.
Ferric Sulphide.	Sesquisulphide of Iron.
Ferric Disulphide.	Bisulphide of Iron.
Diferrous Sulphide.	Subsulphide of Iron.
Cuprous Chloride.	Chloride of Copper.
Cuprous Arsenide.	Arsenide of Copper.
Argentic Chloride.	Chloride of Silver.
Auric Chloride.	Chloride of Gold.
Auric Trichloride.	Perchloride of Gold.

In the names of these neutral binary compounds, we may notice, that the name of the metal *follows* the other name in the old system while it is changed to an adjective and is *followed by* the other in the new. We notice also that the terminations *ous* and *ic* are used in the new system very often, instead of prefixes in the old. To show for example, that there is a larger proportion of chlorine in one compound of chlorine and iron than in another, the new system calls them fer*rous* chloride and fer*ric* chloride; the old calls them *proto*-chloride and *sesqui*chloride.

NEW NAMES.	OLD NAMES.
Sodic Sulphate.	Sulphate of Soda.
Potassic Nitrate.	Nitrate of Potassa.
Potassic Nitrite.	Nitrite of Potassa,
Sodic Carbonate.	Carbonate of Soda.
Hydro-sodic Carbonate.	Bicarbonate of Soda.
Sodic Hyposulphite.	Hyposulphite of Soda.
Plumbic Sulphate.	Sulphate of Lead.
Plumbic Acetate.	Acetate of Lead.
Cupric Nitrate.	Nitrate of Copper,

In these names of salts we see that the name of the class to which the salt belongs is *preceded* by the name of the metal changed to an adjective, in the new system; *followed* by the name of the metal in the old. We notice also that in the new name—hydro-sodic carbonate, the prefix *hydro* is used, while in the old name of the same substance—bicarbonate of soda, the prefix *bi* is given to the other part of the name. So the compound which in the new system is called hydro-potassic sulphate, is in the old, called bisulphate of potassa.

With these few principles in view the thoughtful student will have little difficulty with the names of inorganic substances according to the old system.

To familiarize him still more with the rules of nomenclature the student will find it a valuable exercise to point out the constituents suggested by the names given in the foregoing lists.

(17.) The composition of compounds and the changes they undergo are represented by symbols and equations.

1. *Symbols of compounds.*—Instead of writing the names of compound bodies in full, a system, of symbols has been adopted. The symbol of a compound is made up of the symbols of its constituents with figures to show the number of combining weights. Instead of writing the statement that water consists of hydrogen and oxygen, two combining weights of the one to one of the other, we may with less trouble write the symbol, H_2O, which teaches the same thing. That nitric acid consists of two parts of hydrogen, two of nitrogen, and six of oxygen is shown by the simple expression $H_2N_2O_6$; or if, as is often done, we represent its composition by the simpler symbol HNO_3, we understand it to consist of one part hydrogen, one of nitrogen, and three of oxygen. When no figure is used, one is understood. These symbols contain much valuable information about the compounds they represent.

They teach:—

1st. The names of the constituents, by the letters they contain.

2d. The number of combining weights of each, by the figures they contain.

3d. The number of combining volumes of gaseous or volatile constituents, also by the figures they contain.

4th. The combining *proportion* of each, which is equal to the combining weight multiplied by the figures used.

The symbol of potassic chlorate, for example, is $KClO_3$.

1st. What are its constituents? Potassium, chlorine, and oxygen.

2d. How many combining weights of each? One of potassium, one of chlorine, three of oxygen.

3d. How many volumes of each? One of potassium, one of chlorine, three of oxygen.

4th What combining *proportion* of each? $K = 39.1$, $Cl = 35.5$, $O = 48$.

EXAMPLES.

1st. Write the symbols of the following compounds.

Sulphuric acid—made of hydrogen two parts, sulphur one part, oxygen four parts.

Sulphuric anhydride—one part sulphur, three of oxygen.

Hydro-sodic sulphate.

Potassic nitrate.

2d. Name the compounds having the following symbols:

$$K_2O, \; KHO, \; KCl, \; K_2SO_4, \; KHSO_4.$$

2. *The changes they undergo.*—Let us study the following beautiful experiment. Into a solution of potassic iodide (iodide of potassium) put a few drops of mercuric chloride (corrosive sublimate): almost upon the instant a rich yellow precipitate appears, whose color gradually changes to a bright scarlet. This scarlet-colored precipitate is mercuric iodide. At the same time another substance, potassic chloride, is made, but this being colorless and soluble is not seen.

Now these changes may be shown at a glance by using the symbols of the compounds. Thus:

$$2KI + HgCl_2 = 2KCl + HgI_2$$

Potassic iodide and Mercuric chloride become Potassic chloride and Mercuric iodide.

To understand this, however, we must know that a figure put before the symbol of a compound shows how many combining weights *of the compound* are used. 2KI means two of potassic iodide. The substances *used* in the experiment, form the first member of the equation: the substances *produced* form the second—signs being used as in algebra.

Not an atom is lost, nor an atom gained in these exchanges. The second member must show the same number of combining weights of every element found in the first, but differently arranged to make the symbols of the new compounds. The precision of exchanges is, beyond comparison, perfect. To illustrate this let us write the same equation with the combining weights of the elements (see table p. 21).

$$2KI + HgCl_2 = 2KCl + HgI_2$$
$$2(39.1 + 127) + (200 + 35.5 \times 2) = 2(39.1 + 35.5) + (200 + 127 \times 2)$$

It must be seen that the values of the two members of the equation are exactly equal.

This symbolic language of Chemistry is of the greatest value. It shows, at a glance, an amount of information which if spread out in ordinary language, would often be tedious or obscure, and reveals relations which in the ordinary language would be unseen. Let us become more familiar with the system and with the relations it reveals, and we shall find that among the molecules, as among the planets, an inflexible law prevails, so that the growing, budding, blossoming, and ripening of fruits and the falling of leaves are brought about by as definite laws as those which control the motions of heavenly bodies by which the change of the seasons is produced.

a. *Reaction of sodium and water.*—Drop a bit of sodium upon water: the metal melts, runs about, and gradually disappears: sodic hydrate is formed. The reaction or change is represented thus:

$$\underset{\text{Water.}}{H_2O} + \underset{\text{Sodium.}}{Na} = \underset{\text{Sodic Hydrate.}}{HNaO} + \underset{\text{Hydrogen.}}{H}$$

The equation teaches, at a glance, that one combining weight of sodium takes the place of one of hydrogen in water, forming sodic hydrate, the one combining weight of hydrogen being set free.

Now write the numerical values of the symbols, that is, the combining weights of the substances, and from

$$\underset{18}{H_2O} + \underset{23}{Na} = \underset{40}{HNaO} + \underset{1}{H}$$

we learn that for every 18 grammes, or other units, of water decomposed, 23 of the same units of sodium will disappear, while 40 of the sodic hydrate will be formed and 1 of hydrogen gas will be given off. No human power can change these proportions.

If we would know how much sodium is needed to make 120 grammes of the hydrate, we have but to notice that 40 grammes would need 23 of sodium and then, of course, 120 would need $\frac{120}{40} \times 23 = 69$.

b. *Reaction in the preparation of oxygen.*—We remember that oxygen gas is prepared by heating potassic chlorate: let us examine the chemical changes which take place. Here is the equation:

$$KClO_3 = KCl + 3O$$

Potassic chlorate. Potassic chloride. Oxygen.

This equation shows that the chlorate is decomposed and potassic chloride formed, while 3 combining weights of oxygen are set free.

The numcrical equation, showing the combining weights of the substances, is as follows:

$$\underset{(39.1 + 35.5 + 48)}{K \quad Cl \quad O_3} = \underset{(39.1 + 35.5)}{K \quad Cl} + \underset{48}{3O}$$

By adding the combining weights in the first member we get 122.6. Now 122.6 grammes, or other units of the chlorate will, as the equation shows, give exactly 48 of the same units of oxygen. If we would know how much oxygen could be obtained from any other weight of chlorate—say 613 grammes—we would have, $\frac{613}{122.6} \times 48 = 240$ grammes. Or in general terms we would *divide the given weight of the compound by its combining weight, and multiply the quotient by the weight of the constituent given off, as shown by the equation which represents the reaction.*

By this rule the following problems may be solved.

1. What weight of oxygen may be obtained from 10 grammes of potassic chlorate?

$$\frac{10}{122.6} \times 48 = 3.91 \text{ grammes.}$$

2. How much potassic chloride would be formed?

$$\frac{10}{122.6} \times 74.6 = 6.08 \text{ grammes.}$$

3. How much oxygen by weight in 500 grs. of potassic chlorate? *Ans.* 195.75 grs.

4. How much oxygen by weight in 90 grs. of water?

$$H_2O = H_2 + O$$
$$18 = 2 + 16$$

Then $\frac{90}{18} \times 16 = 80$ *Ans.* 80 grs.

Now observe that, in this equation,

90 is the weight of a compound; let us represent it by C.

18 is its combining weight; let us represent it by c.

16 is the weight of substance obtained from c; let us represent it by a.

80 is the weight of substance obtained from C; let us represent it by W.

And hence $W = a \times \frac{C}{c}$

This formula is a short-hand expression of the preceding rule, and may be easily used to solve problems by substituting given values for the letters, and reducing the equation. For example, how much by weight of oxygen can be obtained from 367.8 grammes of potassic chlorate.

The value of C, is given = 367.8
" " c is known = 122.6
" " a is " = 48

The last two values are taken from the equation which shows the reaction. By putting these values for the letters in the formula it becomes

$$W = 48 \times \frac{367.8}{122.6}; \text{ hence } W = 144. \textit{ Ans.}$$

The superior value of a formula like this is clear when we notice that it contains *four* different quantities, any three of which being given the fourth may be found, so that it may be used to solve *four classes* of problems instead of one.* The following examples will illustrate.

1. How much oxygen can be made from 490.4 grs. of potassic chlorate?

Here C = 490.4; c = 122.6; a = 48; and W is required. W = 192 grs.

2. How much potassic chlorate would be needed to give 96 grs. of oxygen?

Here W = 96; c = 122.6; a = 48; and C is required. C = 245.2.

3. What is the combining weight of potassic chlorate if 100 grs. of it will yield 39.13 grs. of oxygen?

Here C = 100; W = 39.13; a = 48; and c is required. c = 122.6.

4. How much oxygen in the combining weight of potassic chlorate, if 613 grs. of chlorate yield 240 grs. of oxygen?

Here C = 613; W = 240; c = 122.6; and a is required. a = 48.

A different thing being required in each one of these problems, the application of the rule might not be obvious, but substitution in the formula is easy.

c. *Reaction in preparing hydrogen.*—When hydrogen is obtained by the use of zinc and sulphuric acid, a chemical action occurs, represented by the following equation.

$$Zn + H_2SO_4 = ZnSO_4 + 2H.$$

The zinc takes the place of the hydrogen in combination with the acid; zincic sulphate is formed and hydro-

* See report of the Regents of the University of the State of New York, 1867, pp. 603 to 617.

gen gas set free. The numerical equation, by putting combining weights in place of symbols, is:

$$65.2 + 98 = 161.2 + 2$$

which shows that for every 65.2 grammes, or other units of zinc, 2 of hydrogen gas will be set free.

The following problems may be solved:

1. How much hydrogen may be obtained by the action of 32.6 grammes of zinc on sulphuric acid?

2. How much sulphuric acid would be needed?

3. How much zincic sulphate would be formed?

4. How much zinc needed to prepare 5 grammes (77.15 grs.) of hydrogen? *Ans.* $\frac{5}{2} \times 65.2 = 163$ grammes, or $\frac{77.15}{2} \times 65.2 = 2515$ grs.

5. How much zinc required to give 51 grs. of hydrogen.

d. *From weight to calculate volume.*—We pass now to another point of much importance. We have just seen how to calculate the weight of oxygen and other gases obtained by chemical reactions: but gases are to be *measured*, not weighed. Can we from these weights obtain the *volume* of gas set free?

For this purpose we need to remember that, at the standard pressure and temperature, one liter (61.03 cub. in.) of air weighs 1.2932 grammes (19.95 grs.), and that this, multiplied by the specific gravity of any gas will give the weight of 1 liter of it.

For example, the specific gravity of oxygen is 1.105, and the weight of 1 liter (61.03 cub. in.) must be $1.2932 \times 1.105 = 1.429$ grammes. Now, any given weight of this gas, divided by the weight of one liter, will, of course, show the number of liters.

1. How many liters of oxygen in 75 grammes?

1st. $1.2932 \times 1.105 = 1.429$.

2d. $\frac{75}{1.429} = 52.48$ liters.

2. How many liters of hydrogen in 75 grammes?

1st. $1.2932 \times .0692 = .0894$.

2d. $\frac{75}{.0894} = 839.92$.

3. How many liters of oxygen may be obtained from 61.8 grammes of potassic chlorate? *Ans.* 16.79.

4. How many liters of hydrogen may be obtained by the action of 163 grammes of zinc upon sulphuric acid? *Ans.* 55.92.

In the foregoing method we have supposed air to be the standard or unit of specific gravity; but, as we have learned, hydrogen is the unit adopted by chemists. How, then, shall we calculate the volume from the weight of a gas? The *operation* is the same as before; the *values* employed are different.

The weight of one liter of hydrogen at 0° C and $0^{m}.76$ pressure is 0.08936 (Roscoe). The specific gravity of an elementary gas is expressed by the same number as its combining weight; that of a compound gas by one-half its combining weight. The weight of a liter of hydrogen, multiplied by the specific gravity of a gas, gives the weight of one liter of the gas.

1. Let us now calculate the volume of nitrogen gas whose weight is 75 grammes.

1st. $0.08936 \times 14 = 1.251$.

2d. $\frac{75}{1.251} = 59.95$ liters.

2. How many liters of hydrogen may be obtained from 50 grammes of zinc, with sulphuric acid enough to use that amount?

$$\underset{65.2}{Zn} + \underset{98}{H_2SO_4} = \underset{161.2}{ZnSO_4} + \underset{2}{2H}.$$

Then $\frac{50}{65.2} \times 2 = 1.533 =$ grm. of H. from 50 grm. of Zn.

and $\frac{1.533}{0.08936} = 17.155 =$ liters of H. from 50 grm. of Zn.

3. How many liters of oxygen may be obtained from 60 grammes of potassic chlorate?

4. From 100 grammes of ammonic nitrate, how much nitrous oxide may be obtained?

5. What volumes of hydrogen and oxygen would be set free by decomposing 40 grammes of water?

In making these calculations, we have supposed the temperature to be 0° C. or 32° F.; at higher temperatures the volume would of course be greater. It has been found by most careful experiments that gases expand $\frac{1}{273}$ of their volume at 0° C. for every 1° of heat applied to them. Thus, at 15° C., the volume will be $\frac{15}{273}$ greater than at 0° C. Having first found the volume at 0° C., it is therefore very easy to calculate it at any higher temperature.

Let the student apply this to the foregoing problems, and find the volume of the gas at 20° C. in each case.

If the volume is to be calculated according to *Fah-*

renheit's scale, then remember that gases expand $\frac{1}{490}$ of their volume at 32° F. for each additional degree.

Let the student apply this to the foregoing problems, and calculate the volume of the gas at 70° F. in each case.

e. *Reaction in preparing nitrous oxide.*—Let us now study the process of getting nitrous oxide from ammonic nitrate (see p. 53). The symbol of the nitrate is NH_4NO_3, and the reaction is as follows:—

$$NH_4NO_3 = 2H_2O + N_2O.$$
$$14 + 4 + 14 + 48 = 2(2 + 16) + 28 + 16.$$

Now to this reaction apply the methods just illustrated, and make the following calculations:—

1. From 50 grammes of the nitrate, how many liters of nitrous oxide (sp. gr. 1.527) may be obtained? *Ans.* 13.56.

2. How many grammes of water would be set free? *Ans.* 21.94+.

3. From 10 grammes of the nitrate, how many liters of nitrous oxide may be obtained? *Ans.* 2.71+.

4. How much nitrate needed, to give 2.71 liters of nitrous oxide?

$1.2932 \times 1.527 \times 2.71 = 5.35$ = weight of 2.71 liters of oxide. $\frac{5.35}{44} \times 80 = 9.72+$ grammes.

5. How much nitrate needed, to give 100 cub. in. of nitrous oxide?

(18.) From the symbol of a compound we may calculate its percentage composition if we know the combining weights of its constituents. By examining the process we may make a formula by which three other classes of problems may be solved.

1. *Percentage composition.*—By percentage composition we mean the weights of the constituents in 100 parts of a compound. To illustrate: when water is analyzed the chemist finds that every 100 parts of it contain 11.11 parts of hydrogen and 88.89 parts of oxygen. The percentage composition of water is, therefore, 11.11 of hydrogen and 88.89 of oxygen. The percentage composition may be found by analysis of the compound.

2. *Calculated from the symbol.*—Or it may be calculated from the symbol, if we know already the combining weights of its elements. For example, the symbol for acetic acid is $C_2 H_4 O_2$ and we know the combining weights C = 12, H = 1, O = 16.

In the first place multiply each combining weight by the number of atoms of its constituent, the sum of the products we know to be the combining weight of the compound, and each product is the weight of its element in this combining weight. Thus:

$12 \times 2 = 24 =$	weight of	C	in one combining	weight	of acetic	acid.
$1 \times 4 = 4 =$	"	H	"	"	"	"
$16 \times 2 = 32 =$	"	O	"	"	"	"
$60 =$	....	one	"	"	"	"

Being 24 of C. in 60 parts, there are $\frac{24}{60} \times 100 = 40$ in 100 parts.

" 4 " H. " " $\frac{4}{60} \times 100 = 6.66$ "

" 32 " O. " " $\frac{32}{60} \times 100 = 53.34$ "

Hence the percentage composition of acetic acid is C = 40, H = 6.66, and O = 53.34.

3. *Make a formula.*—To obtain the 40 of carbon what has been done? Observe that the operations are as follows:

$$12 \times 2 \div 60 \times 100 = 40.$$

12 is the combining weight of the element; let us represent it by A.

2 is the relative number of its atoms; let us represent it by N.

60 is the combining weight of the compound; let us represent it by E.

40 is the percentage part of the element; let us represent it by P.

Hence $\frac{A\ N}{E} \times 100 = P$.

By a little attention we will see that this formula shows the operations by which the percentage parts of hydrogen and oxygen, as well as that of carbon, were obtained. By applying it to the several elements shown in a symbol, the percentage composition of the compound may be obtained.

What is the percentage composition of alcohol whose symbol is $C_2\,H_6\,O$, and combining weight = 46?

Here for C, A = 12, N = 2; hence $\frac{12 \times 2}{46} \times 100 = P = 52.18$, C.

" H, A = 1, N = 6; " $\frac{1 \times 6}{46} \times 100 = P = 13.04$, H. *Ans.*

" O, A = 16, N = 1; " $\frac{16 \times 1}{46} \times 100 = P = 34.78$, O.

4. *Three other classes of problems.*—By looking again at the formula we may see that it contains four variable quantities, any three of which being given the fourth may be found. Hence *four* classes of problems may be solved :

1st. To find the percentage composition = P.

2d. To find the relative number of atoms of the constituents = N.

3d. To find the combining weight of the compound = E.

4th. To find the combining weight of constituents = A.

The first has been already illustrated: we pass to the second.

Suppose that by analysis the percentage composition of an acid has been found to be C = 40; H = 6.66; O = 53.34, and its combining weight = 60. The combining weights of the elements being known, *what is the symbol of the acid?* Here for C, P = 40; A = 12; and E is 60. Since to find the symbol we must find the relative *number of atoms*, N is required. Putting these values in the formula :—

For C $\frac{12 \times N}{60} \times 100 = 40$; hence N = 2 = number of C atoms.

For H $\frac{1 \times N}{60} \times 100 = 6.66$; hence N = 4 = " H "

For O $\frac{16 \times N}{60} \times 100 = 53.34$; hence N = 2 = " O "

Writing the symbols of the elements with the numbers of atoms we have the required symbol $C_2 H_4 O_2$.

The combining weight of the compound may be found by the formula, if we know the combining weight of one constituent, the relative number of its atoms, and the percentage part of it in the compound.

Let us suppose that the combining weight of ethyl is to be found. For this purpose ethylic iodide is analyzed, and found to yield, in 100 parts, 81.4 of iodine, whose combining weight is 127. The relative number of atoms of iodine is supposed to be 1.

Here $P = 81.4$; $A = 127$; $N = 1$. By putting these values in the formula we have $\frac{127 \times 1}{E} \times 100 = 81.4$; hence $E = 156$. This 156 is the combining weight of *ethylic iodide*, and of course the sum of those of iodine and ethyl. Hence $156 - 127 = 29 =$ the required combining weight of ethyl.

The combining weight of a constituent may be found by the same formula, if we know the relative number of its atoms, the percentage part in a compound, and the combining weight of the compound.

Suppose that an analysis of plumbic acetate has shown the combining weight of acetic acid to be 60; that an analysis of the acid has shown that 100 parts yield 40 of carbon. The relative number of carbon atoms in the acid being 2, what is the combining weight of carbon. *Ans.* 12.

In this case $E = 60$; $P = 40$; $N = 2$. By putting these values into the formula we have $\frac{A \times 2}{60} \times 100 = 40$; hence $A = 12$.

The algebraic expression of a rule "shows at a glance *all the relations* of the data considered, and provides *obvious* solutions for a great variety of problems." *

* See Proceedings of the University Convocation of the State of New York, August, 1866, published in the Regents' Report, 1867.

CHAPTER III.

OF CHEMICAL GROUPS.

General Statement.—The substances to be described in chemistry are so numerous that, to be studied successfully, they must be arranged in groups. No system of classification is yet in all respects perfect. But when our object is to become acquainted with the properties of substances, that system is best which brings into the same group bodies whose properties are most nearly alike.

I.—The Non-Metals.

(19.) In the study of the non-metals, a system of classification, based upon *quantivalence*, is most advantageous.

1. *Quantivalence.*—Let us approach this subject by comparing the composition of hydrochloric acid, water, and ammonia. For this purpose notice their symbols:—

$H\,Cl$	$H_2\,O$	$H_3\,N$
Hydrochloric Acid.	Water.	Ammonia.

And observe that one combining weight of chlorine takes *one* of hydrogen; while one of oxygen takes *two*, and one of nitrogen takes *three* of hydrogen. These particular elements are not at present known to combine with any *larger* proportions of hydrogen.

Not only can one atom of chlorine take no more than

one of hydrogen, but when substitution occurs, one atom of chlorine can *take the place of* only one of hydrogen. Thus:—

$$\text{Cl} + \text{H H O} = \text{H} + \text{H Cl O}.$$

Chlorine + Water = Hydrogen + Hypochlorous Acid.

It would, therefore, seem that in combination, *one* atom of chlorine is equivalent to one atom of hydrogen. Now, to show this peculiarity, chlorine is called a *univalent* element.

One atom of oxygen can hold *two* atoms of hydrogen in combination, or may be substituted for two of hydrogen: on this account it is called a *bivalent* element.

And since one atom of nitrogen in combination seems to be equivalent to *three* of hydrogen, nitrogen is called a *trivalent* element.

But these three elements are types of as many groups of bodies. All substances which, like chlorine, can replace hydrogen, atom for atom, are univalent; those which, like oxygen, can be substituted, one atom for two of hydrogen are bivalent; and those which, like nitrogen, may be substituted one atom for three of hydrogen, are trivalent. In addition to these three we may notice a fourth group, of which one atom of any member may be substituted for *four* atoms of hydrogen: these are *quadrivalent.*

By the term quantivalence, then, we understand the combining power of a substance as measured by the number of hydrogen atoms, which one of its atoms is able, most generally, to replace.

The quantivalence of a substance is measured by the number of hydrogen atoms, which one of its atoms *most generally* replaces: this number is not always the *greatest* number which it *can* replace. "Each element, however, has a maximum power which it never exceeds; this we shall call its *atomicity*, and we shall distinguish the elements as *monads*, *dyads*, *triads*, and *tetrads*, according to the number of univalent atoms they are able at most to bind together." (Cooke's Chem. Phil.)

2. *Mutual quantivalence.*—One atom of any univalent element may replace one of another; but it will take two of its atoms to replace one of any bivalent element; three to replace one of any trivalent element, and four to replace one of any quadrivalent element.

One atom of a bivalent element can replace two of any univalent, and but one of any bivalent element. It will take *three* of its atoms to replace *two* of a *trivalent* element, and two of its atoms to replace one of any quadrivalent element.

Let us represent the quantivalence of the elements in the usual way—*by Roman numerals, written above the symbols*, and we have:—

$\overset{\text{I}}{\text{Cl}}$ . . . One atom of univalent chlorine.

$\overset{\text{II}}{\text{O}}$ . . . One atom of bivalent oxygen.

$\overset{\text{III}}{\text{N}}$ . . . One atom of trivalent nitrogen.

$\overset{\text{IV}}{\text{C}}$ . . . One atom of quadrivalent carbon.

These four symbols represent types of the four groups named.

Now, to show that two univalent atoms of chlorine are required to replace one bivalent atom of oxygen, we may write:—

$$2\,\overset{\text{I}}{\text{Cl}} = 1\,\overset{\text{II}}{\text{O}}.$$

In the same way:

$$3 \overset{II}{O} = 2 \overset{III}{N}$$

shows that 3 bivalent atoms of oxygen are equivalent to, or may replace two trivalent atoms of nitrogen in combination. Observe: The *Roman numeral* of each element shows the *number of atoms* of the other. This will always be the case when the quantivalence of two elements in combination is just balanced.

To apply this principle to carbon and oxygen: how many atoms of each of these elements are equivalent in combination? Making the value of each Roman numeral the co-efficient of the other symbol, we have:—

$$2 \overset{IV}{C} = 4 \overset{II}{O}.$$

But, since the ratio is the same, we may as well take the smaller numbers, and have:—

$$1 \overset{IV}{C} = 2 \overset{II}{O},$$

that is, one quadrivalent atom of carbon is able to replace two bivalent atoms of oxygen. The *products of co-efficient and Roman numerals* must, in all cases, be *equal for the two elements.*

In a great many compounds the quantivalence of the elements is not balanced. In nitrous oxide, for example, $\overset{III}{N}_2 \overset{II}{O}$, we notice, that of the nitrogen there are III × 2 = 6 units of quantivalence, while of the oxygen there are but II × 1 = 2 units of quantivalence. Four of the six units of the nitrogen are unsatisfied. In nitric anhydride, $\overset{III}{N}_2 \overset{II}{O}_5$, we notice, that of nitrogen there are III × 2 = 6 units of quantivalence, while of oxygen there are II × 5 = 10 units of quantivalence. In this case, four of the six units of oxygen are unsatisfied. In nitrous anhydride, $\overset{III}{N}_2 \overset{II}{O}_3$, we see, that of nitrogen there are $III \times 2 = 6$ units of quantivalence; and of oxygen there are $II \times 3 = 6$ units also. All the units of quantivalence in this compound are satisfied. Of the five different compounds of nitrogen and oxygen, the nitrous anhydride is the only one in which this relation exists. When all the units of quantivalence are satisfied, the compound is said to be *saturated.*

3. *Quantivalence of compounds.*—Many compounds also belong to the groups already described. Some will replace one atom of hydrogen, others two; the first are univalent compounds, the last bivalent. Others may be trivalent or quadrivalent.

The compound $N O_2$ is univalent; for,

$$\underset{\text{Water.}}{\left.\begin{matrix} H \\ H \end{matrix}\right\}} O + N O_2 = \left.\begin{matrix} H \\ N O_2 \end{matrix}\right\} O = \underset{\text{Nitric Acid.}}{H N O_3}$$

we see that one molecule of $N O_2$ takes the place of one atom of hydrogen in water and forms nitric acid.

4. *Quantivalence a basis of classification.*—The quantivalence of the *elements* is for the most part very well established, and it is found that those which belong to the same group are, in general, alike in other chemical, and in physical properties. This is more especially true of the non-metals.

I.—THE UNIVALENT NON-METALS, OR CHLORINE GROUP.

(20.) Chlorine is a very abundant element. It is most easily obtained from hydrochloric acid. It is a greenish yellow gas, having a strong affinity for hydrogen and the metals.

1. *Chlorine in nature.*—Chlorine is not found free in nature, but in combination it is one of the most abundant elements. Sodic chloride (common salt, $Na\,Cl$) is distributed throughout the air, the soil, the rocks, and the sea. More than half $\left(\frac{35.5}{58.5}\right)$ the weight of this substance is chlorine, and calculating from the amount of salt in sea-water, we may find that something more than five gallons of this gas is contained in one gallon of sea-water.

2. *Obtained from hydrochloric acid.*—The chlorine in hydrochloric acid may be set free by the action of manganic dioxide. For this purpose, the acid, with about one-third of its weight of the dioxide, is put into a large flask (Fig. 26), provided with a tight cork and

Fig. 26.

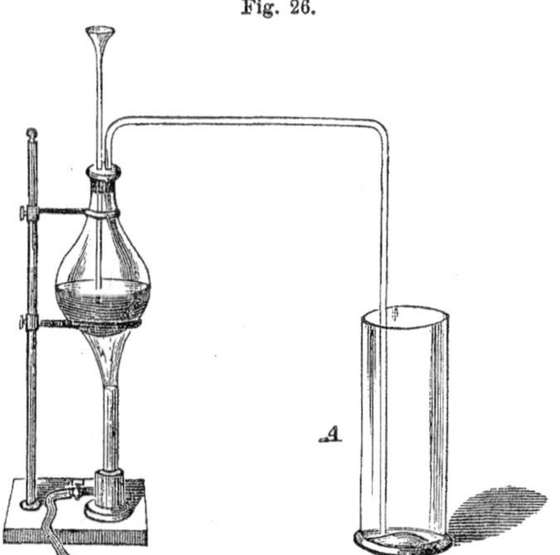

suitable tubes. By heating the mixture gently, the chlorine is given off in abundance; passes over through the tube to the bottom of the jar, A, above which it gradually rises, until the jar is filled.

The reaction is as follows:—

$$Mn\,O_2+4\,H\,Cl=2\,H_2\,O+Mn\,Cl_2+2\,Cl.$$

The oxygen of the dioxide, combining with the hydrogen of the acid, forms water; while a part of the chlorine, taking the manganese, forms manganic chloride, and the rest of the gas is set free.

3. *Its physical properties.*—Unlike other elementary gases, chlorine has a greenish-yellow color, and a peculiarly suffocating odor. It is very soluble in water; for this reason it is not conveniently collected over water, as the other elementary gases may be; but, being about two and a half times heavier than air (2.47), it is easily collected by *displacement* of air. (See Fig. 26.)

4. *Its chemical actions.*—Chlorine combines readily with hydrogen. When a mixture of the two is placed in the direct rays of the sun, a violent explosion announces their union. Diffuse light causes the combination gradually, while, if mixed and kept in the dark, no chemical action takes place.

So strong is the chemical force, or affinity, between these elements that chlorine will decompose many compounds of hydrogen. An instructive experiment illustrates this. A lighted wax-taper, plunged into a jar of chlorine, is extinguished; but, curiously enough, it is at once relighted, burning afterward with a dark, red flame, and giving off a dense, black smoke. The explanation is this:—the white flame in the air is due to the action of oxygen; there being none of this element in the jar, the action ceases. The wax contains hydrogen, and the chlorine, decomposing the wax, combines with this element so vigorously as to produce a flame.

Chlorine is largely used in the arts of bleaching and disinfecting. Its power to destroy colors and bad odors is due to its affinity for hydrogen. It takes this element out of the coloring matter, and at the same time decomposes the water which is present, liberating oxygen, which also attacks the colored substance. The reaction is complicated; but doubtless both chlorine and the oxygen set free by it, destroy the color or the odor, as the case may be, by forming colorless or odorless compounds with its elements.

Chlorine combines, also, with most metals readily. Powdered antimony sprinkled into a tall jar of chlorine, takes fire and falls to the bottom in a shower of sparks. Metals which, like gold, resist the action of oxygen and the acids, are attacked by chlorine. Neither nitric acid or hydrochloric acid alone will act upon gold, but when mixed they form what is called *aqua regia*, in which there is free chlorine; by this mixture gold tri-chloride is speedily formed.

(21.) Bromine is found in small quantities combined with metals in the waters of the sea. It is a dark brown-red liquid, having a strong affinity for hydrogen and the metals. It is used to some extent in medicine, and quite largely in photography.

(22.) Iodine is found in sea-water in still smaller quantities than bromine. It is a blue-black crystalline solid, very volatile, giving off a superb violet-colored vapor when warmed. It has a strong affinity for hydrogen and metals. It is used in photography, and is highly prized in medicine.

(23.) Fluorine is found combined with metals in certain minerals—fluor-spar (calcic fluoride, $Ca Fl_2$) being the most common. It is obtained free only with the greatest difficulty, and on this account its physical properties are imperfectly known. It seems to be a gas, having a violent affinity for hydrogen, for the metals, and indeed for most other substances.

(24.) The properties of these four elements serve to rank them together in a well-marked natural group.

1. *Their physical properties.*—Leaving fluorine out of the account, we notice that chlorine is a gas at ordinary temperatures, bromine a liquid, and iodine a solid. At a little higher temperature all three are gaseous. In the colors of these gases we notice a curious gradation. Bromine is at one end of the spectrum, dark red; chlorine in the middle part, greenish yellow; and iodine at the other end, a beautiful violet.

2. *Their chemical properties.*—These four elements resemble each other in their chemical properties also. They combine with the same substances, and generally in the same proportions. For hydrogen and the metals they all have strong attraction; with oxygen they (with the possible exception of fluorine) unite with a feeble force; with nitrogen they form explosive compounds. They are all univalent, and the single compound which each one forms with hydrogen is an acid.

The following symbols illustrate the analogous composition of the compounds of these elements:

With Hydrogen.	With Oxygen.				
HCl,	Cl_2O_7	Cl_2O_5	ClO_2	Cl_2O_3	Cl_2O.
HBr,	....	Br_2O_5	....	....	Br_2O
HI,	I_2O_7	I_2O_5	....	....	....
HFl,	....	....	....	....	....

1. What names shall be given the compounds of hydrogen with these elements?
2. What number of atoms of each element in these acids?
3. What proportions by volume?
4. What proportions by weight?

The compounds of these elements with oxygen, whose symbols are given, are not all known in a free state; most of them are found combined with water, and in this case they are acids. The Cl_2O_7 is then changed to $HClO_4$, which is called perchloric acid.

$$Cl_2O_7 + H_2O = H_2Cl_2O_8 = 2\ HClO_4.$$

1. Name the other acids corresponding to the several oxides represented.
2. What would be the symbol for each?
3. Describe the composition of each acid: first, by atoms; second, by weight; and third, by volume.

II.—THE BIVALENT NON-METALS, OR SULPHUR GROUP.

(25.) Sulphur is a very abundant element in nature. It is sometimes found crystallized. In commerce it is obtained chiefly in two forms,—flowers of sulphur, and roll brimstone. It is used very extensively in the arts.

1. *Sulphur in nature.*—In some volcanic districts sulphur is found free. The mines of sulphur on the

island of Sicily contain the sulphur mixed with earthy matter, and much more rarely in the form of very pure and beautiful crystals.

Sulphur, in combination with metals, is found in the earth almost everywhere. Iron pyrites, so common and familiar, is a disulphide ($Fe\ S_2$). In combination with hydrogen it is found in the water of what are called sulphur springs. And besides all this, sulphur is an important element in many animal and vegetable substances.

2. *Sulphur in commerce.*—By far the greater part of sulphur in commerce has been obtained from the "native sulphur," such as is found in Sicily, simply by the application of heat. The sulphur is vaporized, and, passing over into large, cold chambers, the vapor is again condensed. In this way sulphur, in the finest powder, and known in commerce as *flowers of sulphur*, is obtained.

When smaller chambers are used, their walls soon become so heated by the hot vapors that the sulphur is kept in a melted state. This liquid, drawn off into molds and cooled, forms what is known as *roll brimstone.*

3. *Its physical properties.*—Many of the physical properties of sulphur are too familiar to need a statement here; the effects of heat upon it, however, are peculiar and worthy of particular notice.

Let a small quantity of sulphur be laid on a piece of writing paper and held over the flame of a candle; in a little time, it melts to a clear, yellow liquid. The melting begins at about 110° C. (230° F.). Again: put more of it into a test tube and heat it gradually. After melting, it remains a clear, limpid liquid up to about

132° C. (270° F.), and then begins to get thick and dark colored. At about 260° C. (500° F.) it is so viscid that it can scarcely be poured from the tube; but, as the heat increases, it becomes less viscid, and finally, at about 427° C. (800° F.), it boils.

If, when heated to about 232° C. (450° F.), it is poured into cold water, it becomes curiously unlike common sulphur: it cools into a dark-colored solid, with a considerable degree of *elasticity.*

4. *Its chemical properties.*—Sulphur has a wide range of attractions. It unites with oxygen and with hydrogen to form important acids. With the metals it forms a numerous class of sulphides or, as they were formerly called, *sulphurets.*

5. *Its uses.*—Sulphur is very largely used in the arts. It is a constituent of gunpowder and of sulphuric acid. It is used in the manufacture of friction matches, in medicine also, and, in its elastic state, produced under the influence of heat, it is used in taking casts of coins and medals.

(26.) The other members of the sulphur group are oxygen, which has been already described, and the two rare and unimportant elements, selenium and tellurium.

(27.) The four elements just considered form a second well-marked natural group. Their affinities are for the same substances, and their compounds have a similar composition.

1. *Their compounds are analogous.*—Let us first notice the hydrogen compounds of the members of this group. They are shown by the following symbols:—

$$H_2O,\ H_2S,\ H_2Se,\ H_2Te.$$

One atom of each is able to combine with two of hydrogen. The bivalent character of these elements is clearly shown in these compounds.

The compound of hydrogen and sulphur is worthy of more than a passing notice. Into a bottle (Fig. 27), provided with a very tight cork and proper tubes, some

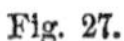

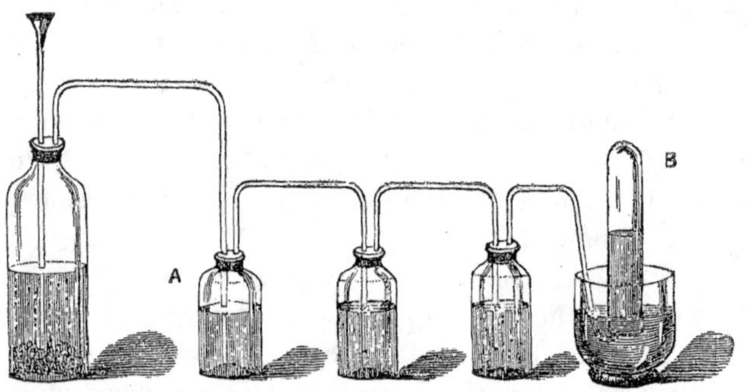

ferrous sulphide (Fe S) is placed, and through the funnel tube is poured dilute sulphuric acid. A colorless gas will pass over through the delivery tube into the water of the first bottle A, of the series. A part of the gas will be dissolved in this water; what is not, will pass over to the next bottle, and, if the operation be continued, the gas will finally escape into the jar B, at the end of the series. There will then be in the bottles a solution of the gas, while the gas itself will be collected in the jar; and, should any escape into the room, its presence would be known by its fetid and even disgusting odor. This gas is the hydrosulphuric acid (H_2 S), commonly called sulphuretted hydrogen—one of the most valuable agents in the laboratory of the chemist. In the work of analysis it is indispensable.

2. *Compounds of oxygen with the other three.*—The last three elements of the group form analogous compounds with oxygen. Notice the symbols:

$S\,O_2$	$Se\,O_2$	$Te\,O_2$
$S\,O_3$	$Se\,O_3$	$Te\,O_3$

in which this resemblance clearly appears.

By combining with water these oxides all form acids: the sulphurous ($H_2\,S\,O_3$) and the sulphuric ($H_2\,S\,O_4$) being very useful substances. The former is used for bleaching straw and woolen goods; the latter, in immense quantities, for the manufacture of chemicals.

III.—THE TRIVALENT NON-METALS, OR NITROGEN GROUP.

(28) Phosphorus, in small proportions, is found in some minerals, and is a most important element in organic bodies. It is obtained from bones. It is a waxy-looking solid, shining in the dark, and having a strong attraction for oxygen.

1. *Phosphorus in nature.*—Compounds containing phosphorus are quite common in soils and rocks. Plants receive them from the soil, while animals, living upon vegetable food, take these compounds from the plant. It is found in wheat and other grains, and in the brain and secretions of animals: so that while not very abundant, it is a widely distributed and very important element. Its most common compounds are the phosphates, and of these the most abundant is the calcic phosphate. This compound of phosphorus occurs in bones, and from these the element is obtained.

2. *Its physical properties.*—Phosphorus is a solid element, having a pale yellow color, rather soft and

wax-like at common temperature, but brittle at 32° F. In a dark room its clean surface shines with a feeble, pearl-white light. Its vapor also is beautifully phosphorescent. To show this, put a few small fragments of the solid into a flask of water and boil it. The mixed vapors of water and phosphorus escape, and, in a dark room, look like livid flames, while at the same time globules of melted phosphorus, on the surfaces of the water and the glass, appear like little balls of pearl. This element is insoluble in water, but it dissolves freely in ether. If this solution be rubbed over the skin of a person in a dark room, his appearance becomes extremely ghost-like, because the ether evaporating leaves the phosphorus, with its peculiar glow.

3. *Its chemical properties.*—Phosphorus combines most readily with oxygen. From a piece of the solid exposed to air, white fumes are continually falling, which consist of a compound of phosphorus and oxygen. A gentle heat—that of the fingers handling it is often enough—causes it to burst into violent combustion, forming the fumes in great abundance. On this account the element must be kept under water, and it should be held under water when cut, lest the friction of the knife set it on fire. The pieces to be used should be afterward dried by gentle pressure between layers of blotting paper. There is, however, an allotropic form of phosphorus, called red phosphorus, which will not produce the phenomena of combustion under a temperature of 260° C. (500° F., Brande & Taylor). Experiments with this variety are, of course, far less dangerous than with the common form. Sticks of the solid, exposed to light, gradually change to red phosphorus.

This element is a violent poison; even its vapo inhaled, will cause wasting disease.

Large quantities of phosphorus are used in the manufacture of friction matches: sulphur is also used. Friction sets fire to the phosphorus, the burning phosphorus inflames the sulphur, and this, in turn, ignites the wood.

(29.) Arsenic is found in nature, chiefly associated with the metals. It is a brittle solid, of a steel-gray color, forming compounds remarkable for their poisonous effects.

1. *Arsenic in nature.*—Arsenic sometimes occurs in the earth uncombined with other substances; but generally it is found as an oxide or a sulphide, mixed with similar compounds of the metals. "It was at one time supposed that arsenic entered into the composition of the flesh and bones of animals as a normal constituent, but it has been clearly proved that it is never found in the tissues, either of animals or vegetables, except when it has been introduced into them by accident or design."—(Brande & Taylor.)

2. *Its physical properties.*—Arsenic is a very brittle solid, in appearance much like metals. It may be sublimed by heat, that is to say, it may be changed from the solid to the vapor state, directly, without melting.

3. *Its chemical properties.*—Heated in the open air, arsenic rapidly combines with oxygen, and even on simple exposure to air it forms an oxide. With some of the metals it forms arsenides. Owing to the criminal use of its poisonous compounds, this element has been most thoroughly and successfully studied. Some points of interest will be given further on.

(30.) Nitrogen, described in an early part of our course, phosphorus, and arsenic are trivalent elements. Other chemical relations also rank them in the same natural group.

1. *They are trivalent.*—The compounds of these elements with hydrogen have the following symbols:—

$$H_3N \qquad H_3P \qquad H_3As,$$

from which we learn that an atom of each can hold three atoms of hydrogen in combination. These three compounds are gases. With the first, ammonia, we are already acquainted. The second, commonly called phosphuretted hydrogen, instantly takes fire, with explosion, on coming in contact with air. It is a dangerous and unimportant compound. The third, commonly called arseniuretted hydrogen, is a combustible gas, to be mentioned when describing tests for the presence of arsenic.

2. *Other chemical relations.*—The complete analogy in the composition of the compounds of these elements may be seen by studying the following symbols (Eliot & Storer, page 285):—

N_2O_3	N_2O_4	N_2O_5	NCl_3 (?)		
P_2O_3	....	P_2O_5	PCl_3	P_2S_3	P_2S_5
As_2O_3	....	As_2O_5	$AsCl_3$	As_2S_3	As_2S_5

The arsenic sesquioxide (As_2O_3) is the common "arsenious acid" or "white arsenic,"—the well-known poison. It is a white solid, slightly soluble in water, with which it forms an acid, and this acid combines with metals to form arsenites. These, also, are violent poisons, and yet some of them are used quite largely in calico printing and as pigments. Green wall-papers, many of them, have been colored with copper arsenite, well known as Scheele's green: they are not to be recommended.

The most minute traces of arsenic poison may be detected by a competent chemist; some of his tests may be studied with interest. (See Eliot & Storer's Chem., p. 252.) That known as Marsh's test is very delicate and sure. It is applied by means of the apparatus shown in figure 28. Into the bottle A are put *pure* zinc, water, and sulphuric acid for the evolution of

Fig. 28.

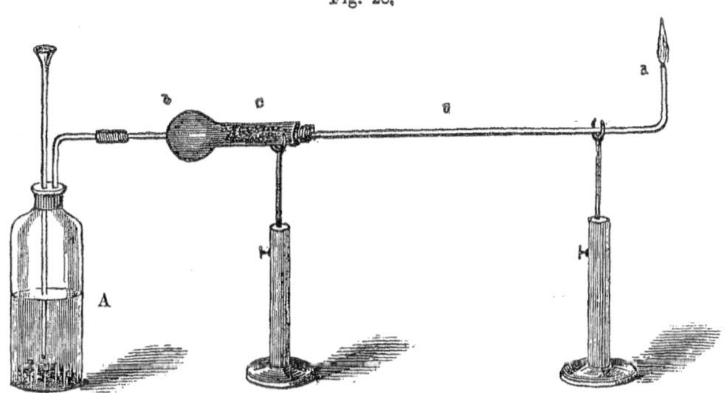

hydrogen gas. The liquid supposed to contain the poison may be afterward poured through the funnel tube. The gas formed in the bottle, passing through the bulb, *b*, loses a part of the water carried over with it, and going over calcic chloride in *c*, it is thoroughly dried. It finally escapes at the pointed end of the tube, *d*. After the air of the apparatus has been all driven out by the stream of gas, the hydrogen may be burned as it issues.

Now, if the liquid contain arsenic, there is at once

formed arseniuretted hydrogen (H_3 As). The color of the flame turns to a livid hue, and sometimes gives off white fumes of arsenious acid. But now comes the decisive test. A cold, clean, white porcelain surface is held in the small flame for a moment; *metallic arsenic* condenses on its surface, if the poison is present, and when cold, appears a blackish-brown stain, with a bright *metallic luster.* The substance of this stain may be still further tested, until the last doubt of its character is removed.

The experiment is varied by applying heat to the middle part, *o*, of the small tube. The arseniuretted hydrogen, if present, will be decomposed and the metallic arsenic will lodge upon the inside surface of the tube, forming a brilliant, mirror-like ring.

This is only one of many tests by which, taken together, the chemist can pronounce upon the presence of arsenic with absolute certainty.

IV.—THE QUADRIVALENT NON-METALS, OR THE CARBON GROUP.

(31.) Silicon, next to oxygen, is the most abundant element in nature. It is a solid, very infusible, and very insoluble. Like carbon it occurs in three allotropic forms.

1. *Silicon.*—This element is found only in combination. With oxygen it forms silicic acid, or, as it is commonly called, silica (Si O_2). This compound is one of the most abundant substances in the earth. The beautiful opal and amethyst and other gems are almost pure silica: so is the more common rock crystal or

quartz, while the sand of the sea-shore and every variety of sandstone rock are the impure forms of the same substance.

Silica is also an important constituent in organic bodies. It gives strength to the stalks of grains and grasses, while it constitutes the skeleton of whole tribes of some of the lower orders of animals.

Silica is used largely in the manufacture of glass.

(32.) Glass is made from silica and the bases, potash, soda, lime, and others, according to the variety required. These materials being intensely heated, melt into a transparent pasty mass, portions of which may be taken from the furnace and blown or molded into the different forms in which glass articles are made. These articles being afterward annealed are ready for the market.

1. *The materials.*—In all true glasses silica is one constituent. This substance exists in almost pure form in flint, agate, and quartz, while all varieties of sand consist of the same in various degrees of purity. Flint was formerly used in the manufacture of glass; hence the name, *flint glass*, which one variety still retains. Sand is now the more general source of silica, great care being used to select a pure material.

Potash, soda, and lime are the most important bases used with the silica; but besides these, plumbic oxide (oxide of lead) and oxides of tin and manganese are often used.

2. *The varieties of glass.*—There are several varieties of glass, among which we will notice four, viz.: green bottle-glass, Bohemian, window-glass, and flint-glass.

Green bottle-glass is made of cheaper and coarser material than any other variety. Its bases are more numerous; oxides of iron and manganese are among them, and to the first of these the glass owes its familiar color.

Bohemian glass is made from purer material than bottle-glass, and care is taken to have it free from color. Its bases are potash, lime, and manganese. Its lightness, the absence of color, and its power to stand high heat and sudden changes of temperature, make it very valuable for chemical purposes.

Window-glass consists chiefly of silica, soda, and lime.

Flint-glass consists chiefly of silica, potassa, and plumbic oxide ($Pb_2 O_3$). This very beautiful variety of glass, sometimes called crystal glass, is chiefly made into such articles of domestic and ornamental use as tumblers, decanters, wine-glasses, and vases. It is very transparent, and refracts light powerfully: on this account it is valuable for lenses, and other optical instruments.

3. *The melting.*—The raw material is heated in pots made of the purest and most infusible clay, and set in a conical furnace. When the heat is sufficient, the silica with the bases form silicates. In the case of window-glass, for example, the silica, soda, and lime form *sodic* and *calcic silicates.* The compound of these silicates is a transparent half-fluid or pasty mass of *glass,* ready to be wrought into any desired form.

4. *The blowing.*—By means of an iron tube four or five feet long, the workman takes out of the furnace a portion of the pasty and adhesive glass. He gives it regular shape by rolling it upon a smooth hard surface,

and makes it hollow by blowing air through the tube. By keeping the tube in constant rotary motion while he blows, the bulb is enlarged into a globe; or if, at the same time, he swings his tube, pendulum like, a pear-shaped flask is made.

Some idea of the interesting process of making window glass may be gained by studying the following cuts found in Muspratt's Applied Chemistry. In Fig.

Fig. 29.

29 some of the first stages of the process are shown. In front of each opening of the furnace is a stage, built over a pit about ten feet deep. Upon these stages the workmen stand. The ball of glass having been taken from the furnace, the workman blows through his pipe,

while he at the same time skillfully rotates it in his hand and swings it backward and forward, sometimes below his feet, sometimes over his head, until he has lengthened and enlarged it into a cylinder with uniform sides as long as the pane of glass is expected to be. The ends of this cylinder are then cut off, and a slit is made along one side. It is afterward placed upon a smooth and even stone plate, and heated in an oven. As the glass softens, the workman, with an iron rod, presses the sides of the cylinder open (see Fig. 30), and finally smooths it out upon the flat surface of the stone. It is then a *pane of glass*, needing only to be carefully cooled.

Fig. 30.

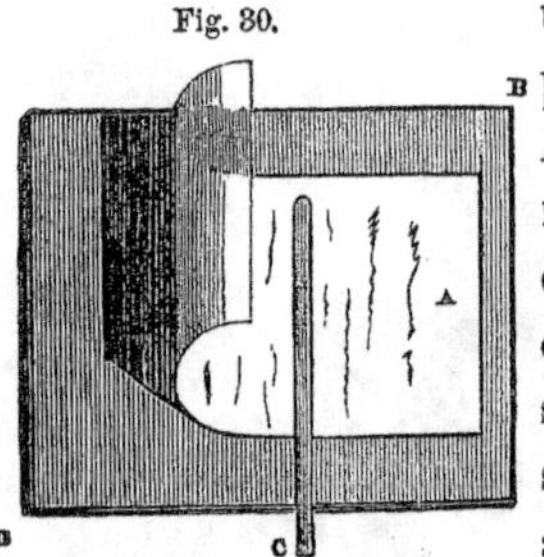

5. *The annealing.*—Annealing is a process of slow cooling. All articles of glass must be annealed. For this purpose they are placed in hot ovens, whose temperature grows gradually less, until, at the end of four or five days, they are quite cold. The process is necessary, because, if glass is cooled suddenly it becomes exceedingly brittle, and will sometimes break even without apparent cause; but, when slowly cooled, it is able to stand much pressure and sudden blows.

"Considered only with reference to its application in the study of natural phenomena, it is impossible to doubt the singular influence glass has exerted on the progress of science. It is chiefly by its aid that astronomy has attained a perfection so wonderful; by it, also, naturalists have been enabled to study under the microscope a host of phenomena which before escaped notice. But perhaps of still greater importance is the use made of it by chemists in their experiments. It requires no profound chemical knowledge to recognize the fact that to glass is chiefly owing the present advanced state of the sciences, so fruitful in marvelous application."

(33.) Carbon, described in an early part of the work, and silicon, are quadrivalent elements. Other properties rank them together as members of the same group.

1. *They are quadrivalent.*—One atom of carbon can hold four of hydrogen in combination: one of silicon can do the same. The symbols of these compounds are $C H_4$ and $Si H_4$. Carbon forms a host of other compounds with hydrogen, some of which will be noticed in due time.

2. *Other properties.* These two elements have each three allotropic forms. The common form of carbon is charcoal: its crystalline forms are graphite and the diamond. So the common form of silicon is a dark-colored solid, which may be changed to two crystalline forms, like graphite and diamond.

Both these elements are very hard, very infusible, and able to resist the action of chemicals in greater degree than most others.

(34.) The compounds of carbon are more numerous than those of all other elements taken together. Many of them are very complex in composition. They are, for the most part, constituents of, or products from, organic bodies. For these reasons they have usually been studied by themselves, in what is called Organic Chemistry.

1. *The compounds of carbon.*—One compound of this element, carbonic dioxide (CO_2), has been described. Another, the carbonic oxide (CO), is a colorless and very poisonous gas. It burns with a blue flame, which most of us have noticed playing over the surface of a freshly made coal fire.

But it is to the compounds of carbon and hydrogen that especial attention should be given.

Marsh gas is the common name of one of these. Who has not seen bubbles of gas rising through the water of stagnant pools? In this case marsh gas has been produced by the decay of dead leaves, or other organic matter, and its name comes from the fact of its occurrence in such places. This same gas collects sometimes in mines, and, by explosions of which we have all heard, now and then extinguishes the lamps and the lives of the miners. On this account it has also been called *fire-damp.* Its symbol is CH_4, and the chemist calls it *methyl hydride.* It is an important constituent of illuminating gas.

Olefiant gas is another compound of carbon and hydrogen, whose composition is shown by its symbol C_2H_4. The chemist calls it ethylene. It burns with a bright flame, being the most luminous constituent of illuminating gas. Mixed with air it explodes violently. Explosions of illuminating gas now and then occur with exceeding violence: life is destroyed and buildings ruined by them.

2. *Are very numerous.*—A multitude of these compounds of carbon are already known, and new ones are all the time being brought to light. That some idea may be formed of this peculiarity, glance at the

following symbols of a part of only three series. (See Eliot & Storer, p. 312.)

Petroleum Series.	Destructive Distillation Series.	Coal Tar Series.
$C H_4$	$C H_2$	$C_6 H_6$
$C_2 H_6$	$C_2 H_4$	$C_7 H_8$
$C_3 H_8$	$C_3 H_6$	$C_8 H_{10}$
$C_4 H_{10}$	$C_4 H_8$	$C_9 H_{12}$
$C_5 A_{12}$	$C_5 H_{10}$	
$C_6 H_{14}$	$C_6 H_{12}$	
$C_7 H_{16}$	$C_7 H_{14}$	

All the members of the first series, together with many others, may be obtained from petroleum. Numerous other series might be named, and their members, together with other compounds formed by the union with them of two other elements, oxygen and nitrogen, may be counted by thousands.

3. *Many of them very complex.*—The molecules of these compounds are often made up of a great number of atoms. In this respect they form a striking contrast with others which we have examined. For example, compare sulphuric acid, $H_2 S O_4$ with the powerful poison strychnine, $C_{21} H_{22} N_2 O_2$. The molecule of the first contains seven atoms, while that of the second contains forty-seven.

4. *Found in organic bodies.*—Plants and animals are *organized bodies.* Their bodies have been nourished and enlarged by means of food taken into and distributed throughout their interior. The plant, by means of certain organs, receives the sap into its roots and sends it to every part. Every leaf and stem is built of substance from this sap, and from the air. The

animal takes its food, converts it into blood, distributes it to the most minute fibers throughout, and every part of its body is built from material thus furnished.

Every distinct part of these bodies is also organized: the leaf, the twig, and the tendril are organized bodies; and so are the hairs, the claws, and the tissues of animals.

These organized bodies consist almost entirely of the compounds of carbon and hydrogen, oxygen and nitrogen. Sugar ($C_{12} H_{22} O_{11}$), starch ($C_6 H_{10} O_5$), and alcohol ($C_2 H_6 O$) are three among the thousands of substances which the chemist is able to get by decomposing organic bodies. Such compounds may be called organic *substances* in distinction from organized *bodies* from which they are obtained.

Now, nature seems to have fixed a dividing barrier, which the chemist may not pass, between organic substances and the organized bodies which they form. The chemist can, by synthesis, make a few of the simplest organic substances, and he has good reason to believe that even the most complex are produced by chemical force governed by the ordinary laws of combination. But, on the other hand, the simplest organized body, although it consists of the same elements, is entirely beyond his reach: he can not make a single cell. By what chemistry the leaf and the flower are made, he does not know. His laboratory teems with elegant crystals whose growth he has himself guided; but his garden is filled with still more delicate forms, the secrets of whose chemistry are known only to the Divine Chemist who made the earth and the sun and all that they contain.

II. The Metals.

(35.) The metals also may be grouped according to quantivalence; but in this case substances would often be thrown together having few points of resemblance. It will be better for our purpose to group them into classes, in which the members have certain general characters in common.

1. *The metals.*—The peculiarities of metallic elements are more or less familiar. Their *luster*, as of silver; their *malleability*, as of gold and zinc; their *ductility*, as of iron and copper; together with their power to conduct heat and electricity, are their most characteristic properties. One or more of these properties are possessed by some non-metals, and, on the other hand, some metals have them only in a slight degree. Indeed, nature seems to have drawn no decided line of division between the two classes. Certain elements, arsenic and antimony for example, are sometimes classed with metals, but oftener now with non-metals; while even hydrogen, because of its chemical relations, is thought by some to be a metal.

The metals, with one exception—mercury, are solids at ordinary temperature. Some are easily melted, as potassium, at 62° 5 C (144° 5 F.); others melt with difficulty, as iron at 1,600° C. (2,912° F.); while others, like platinum, melt only in the intense heat of the oxy-hydrogen blow-pipe.

Iridium is the heaviest of metals, 21.8 times heavier than water; others, as potassium and sodium, will float on the surface of water; lithium being the lightest of all. (Sp. gr. .593.)

A few metals, copper and gold for example, are found in nature in the metallic state: this condition is commonly called *native.* But in the native state they are seldom pure: two or more are combined. Combinations of metals are called *alloys.* They are, however, usually found combined with non-metals, and such compounds are called *ores.*

2. *Their classification.*—In the following table the metals are classed so as to bring together those which most closely resemble each other. Many of the metals are rare and not important to the general student; others are of the greatest use and interest. Of these last the names are printed in capitals, and to the description of them we are to pay the more particular attention.

I.—METALS OF THE ALKALIES.

Potassium,	Cæsium,	Lithium,
Sodium,	Rubidium,	Ammonium.

II.—METALS OF THE ALKALINE EARTHS.

Calcium,	Strontium,	Barium.

III.—METALS OF THE EARTHS.

ALUMINUM,	Zirconium,	Yttrium,
Glucinum,	Thorium,	Erbium,
Cerium,	Lanthanum,	Didymium.

IV.—THE ZINC CLASS.

Magnesium,	ZINC,	Cadmium.

V.—THE IRON CLASS.

IRON,	Nickel,	Uranium,
Manganese,	Chromium,	Indium,
	Cobalt.	

VI.—THE TIN CLASS.

TIN,	Titanium,	Niobium,
	Tantalum.	

VII.—THE TUNGSTEN CLASS.

Molybdenum,	Vanadium,	Tungsten.

VIII.—THE ARSENIC CLASS.

Arsenic,	Antimony,	Bismuth.

IX.—THE LEAD CLASS.

LEAD,		Thallium.

X.—THE SILVER CLASS.

COPPER,	MERCURY,	SILVER.

XI.—THE GOLD CLASS.

GOLD,	Rhutenium,	Osmium,
PLATINUM,	Iridium,	Palladium,
	Rhodium.	

(36.) The metals of the first class are very soft and light, having so violent an attraction for oxygen that they can decompose water at any temperature. They are univalent. The metals themselves are of little use, but many of their compounds are of great value in the arts.

1. *Illustration of these class properties.*—A piece of potassium or of sodium may be molded in the fingers like wax, and if dropped upon water it floats.

Upon a piece of ice place a small fragment of potassium: a purple flame springs up, as if the ice had been set on fire. A smart explosion usually ends the experiment; and if we then examine the water that is left in the cavity of the ice, we find it to contain potassic hydrate (potash). So strong is the attraction of this metal for oxygen that it decomposes water, even at the temperature of ice. This is true of the other members of the group.

If we study the reaction taking place in the experiment, we shall see that the metal is univalent. Thus:—

$$H_2O + K = HKO + H.$$

One combining weight of potassium has simply replaced one of the two combining weights of hydrogen in water. A similar reaction would occur with the other members of the group. The potassic hydrate formed is an *alkali;* the other hydrates of the group are also alkalies. They differ from most other metallic hydrates in their power to withstand heat: heat alone will not decompose them.

(37.) Potassic hydrate is much used in the arts. It is obtained by decomposing potassic carbonate by calcic hydrate. The potassic carbonate is also of much importance in the arts and manufactures: it is obtained by leaching the ashes of plants.

1. *Manufacture of potassic carbonate.*—The ashes of wood, mixed with about five per cent. of lime, are placed in tubs and drenched with successive portions of fresh water. As the water soaks through the ashes, it dissolves out the soluble constituents, among which is the potassic carbonate. The solution, called *lye*, is put into broad and shallow pans, and evaporated. The solid residue is called *crude potash.* By strongly heating this substance, much of its impurities may be driven off: the purer carbonate remaining is called *pearlash.*

A pure salt may be obtained by dissolving the pearlash and then letting it crystallize. The salt crystallizes while the impurities are still in solution: at this point the fluid is drawn off and the crystals left. The symbol of the pure substance is $K_2 C O_3$.

2. *Preparation of potassic hydrate.*—Potassic hydrate (H K O) is obtained by boiling a solution of the carbonate with slaked lime ($H_2 Ca O_2$). A reaction occurs, in which the potassium of the carbonate is substituted for the calcium of the calcic hydrate, or slaked lime, and by this action potassic hydrate is formed, which remains in solution, while the calcic carbonate produced falls to the bottom as a heavy powder.

The clear solution is afterward evaporated to dryness, and the solid hydrate is then fused and run into molds.

Potassic hydrate is very soluble in water, and has a strong affinity for carbonic acid. It greedily takes both these substances from the air, until at length the entire mass of hydrate is changed into a sirup of the carbonate. To indicate this property of dissolving in water absorbed from the air, the term *deliquescence* is used.

(38.) The most familiar compound of sodium is sodic chloride. By the action of sulphuric acid upon this chloride, sodic sulphate is formed, and this, by the action of carbon, is changed to sodic sulphide; and, finally, the sodic sulphide, by the action of calcic carbonate, is changed into sodic carbonate. The production of sodic carbonate is one of the most important branches of chemical manufacture.

1. *Sodic chloride.*—Sodic chloride, so well known as common salt (Na Cl), is everywhere abundant. In many parts of the world it occurs in thick beds, from which it may be mined. Large quantities are obtained

by evaporating the water of salt springs, while immense quantities in solution give its characteristic saltness to the sea.

The uses of this substance are important: not among those least important is its use in the manufacture of the other sodium compounds, especially of sodic carbonate. Enormous quantities of sodic carbonate are used in the arts of bleaching, soap-making, glass-making, and others. The several processes in its manufacture are as follows:—

2. *Sodic chloride changed to sodic sulphate.*—When salt is heated with sulphuric acid in a reverberatory furnace, mutual decomposition takes place, and sodic sulphate with hydrochloric acid are formed. This may be understood by the equation,

$$2\,Na\,Cl + H_2\,S\,O_4 = Na_2\,S\,O_4 + 2\,H\,Cl.$$

The sodic sulphate thus formed is valuable, aside from its use in making the carbonate. It is well known under the name of *Glauber salts.* In the manufacture now being described it is called *salt cake.*

3. *The sodic sulphate changed to sodic sulphide.*—When sodic sulphate is decomposed by carbon, the following reaction occurs:—

$$Na_2\,S\,O_4 + 4\,C = Na_2\,S + 4\,C\,O.$$

Sodic sulphide and carbonic oxide are produced. In the manufacture of sodic carbonate, the sulphate, with small coal and chalk, or limestone (calcic carbonate), are thoroughly mixed and melted together in a furnace. The above reaction takes place.

4. *The sodic sulphide changed to sodic carbonate.*—The sulphide formed by this reaction is at once changed

by the calcic carbonate according to the following equation:

$$Na_2 S + Ca C O_3 = Na_2 C O_3 + Ca S.$$

Sodic carbonate and calcic sulphide are formed. The mixture has a blackish-gray color, and is called *black ash.*

The black ash is afterward thoroughly leached; during this process the water dissolves out the carbonate, but leaves the sulphide; and then, finally, the solution is evaporated to dryness: the residue is the crude sodic carbonate of commerce, generally known as *soda ash.*

(39.) Hydrosodic carbonate (bicarbonate of soda) is obtained by exposing sodic carbonate to the action of carbonic acid. Its symbol is $H Na C O_3$. This is the substance familiarly known as "soda," and used so commonly instead of yeast in bread-making. It is used in medicine: it is also much used in making effervescing drinks.

(40.) The metals of the second class are bivalent. They form carbonates which are not soluble in water, unless it contains carbonic acid: in this respect they differ from the metals of the first class. The metals themselves are of little use, but some of the compounds of calcium and barium are of considerable importance.

1. *Illustrations of these class properties.*—The metals of this class, like those of the first, decompose water at all temperatures, but the reaction is somewhat different. Suppose calcium is used for the purpose, then:

$$H_2 O + Ca = Ca O + H_2.$$

Observe that one combining weight of calcium replaces the two combining weights of hydrogen in water. This illustrates the bivalent character of calcium: the other members of the group are also bivalent.

The calcic oxide (lime), shown in the reaction just given, combines with water to form the calcic hydrate (slaked lime), which is very slightly soluble in water, forming lime-water.

The oxide is made on a large scale, to be used in forming *mortar* and cements, so valuable in building. Limestones are mixed with coal and burned in kilns. The carbonic acid of the limestone is driven off by the heat, and the other constituent, lime, or, as it is often called, *quicklime*, is left, still in the form of hard and compact stone. In contact with water the stone swells, grows intensely hot, and crumbles to powder. The slaked lime thus made is mixed with sand to form *mortar*.

The members of this group form *carbonates*. Limestone and marble of every kind are chiefly the calcic carbonate. These compounds are not soluble in water unless it contains carbonic acid. They then disappear, because they combine with the acid and form hydrocarbonates (*bi*carbonates), which are soluble. The formation of stalactites is a beautiful illustration of this action. Water, charged with carbonic acid, flowing through soil and over rocks where limestone is abundant, dissolves this substance. Finding its way to caverns, it falls drop by drop from the roof. Exposed to the air, carbonic acid evaporates; the water can no longer hold the carbonate in solution, but deposits it wherever it rests. Drop by drop, for a moment clinging to the roof, leaves its mite of carbonate behind, until pendant masses, like icicles, sometimes of curious shape and beauty, are formed. The carbonate left, at the same time, on the bottom of the cave is called stalagmite.

(41.) Aluminum is the most important metal of the third class. The others are rare, even as specimens in a chemist's cabinet, and of no practical importance.

1. *Aluminum.*—This metal has a combination of properties which renders it one of the most interesting in the whole series. It has the hardness and luster of silver, and since it does not tarnish when exposed to air and vegetable acids, it would seem to be fitted for the practical uses to which silver is put. It melts only at a high temperature, and may then be cast into any desired form. This, together with its malleability, ductility, and tenacity, would enable it to replace iron for many purposes, while its lightness (density 2.56) and beauty give it advantage over that metal.

In connection with these valuable properties we find that aluminum is one of the most abundant elements in nature. It is a constituent of clay and marl, of slate, and indeed of most rocks and soils. It must constitute about one-twelfth of the solid parts of the earth.

But no cheap method of extracting the metal is yet known, and the expense stands in the way of its application in the arts. It is now manufactured on a commercial scale in England and France, and it has been used for ornamental work and for optical purposes to some extent.

(42.) Of the fourth class zinc is the most important

metal. The metals of this class are alike fusible at quite low temperatures, volatile at temperatures not exceeding a bright red heat, and combustible when heated in the air. They are bivalent, and form, each, but one oxide, chloride, and sulphide.

1. *Zinc.*—Zinc *blende* (a sulphide), *calamine* (a carbonate), and the red oxide, are the substances containing zinc, found most abundantly, and from these the metal is extracted. When either of the first two are used, it is first *roasted*, that is, heated in presence of air. By this means it is changed to the form of *oxide.* Mixed with coal, the oxide is then heated in a close vessel having an iron tube reaching over into a cold receiver. This oxide is decomposed, and its zinc, in the form of vapor, goes over to be condensed in the receiver.

At low temperature zinc is brittle: heated to about 200° C (392° F.) it is also very brittle; but between these extremes (130° C.) it is malleable, and is rolled or hammered into thin sheets for various uses.

Zinc is not easily acted upon by air, and on this account it is sometimes used as a coating to protect iron from rust. Iron covered with a thin coating of zinc is said to be *galvanized.*

2. *Illustrations of the class properties.*—The melting points of these metals are comparatively low: of zinc at 423° C, of magnesium a little higher, and of cadmium a trifle lower.

Heated to a bright red heat magnesium is changed to vapor; at a low red heat cadmium vaporizes, and zinc at a temperature between these extremes.

At a red heat in the air these metals burn. Cadmium gives the vapors of its oxide; zinc with a blue flame, forming clouds of vapor, and magnesium with a flame of most dazzling brightness, sometimes used as a source of light in photography and in optical experiments.

At high temperatures they decompose water and form oxides. In this reaction one atom of metal replaces two of hydrogen, and forms an oxide with the one atom of oxygen, and hence they are bivalent.

(43.) Iron is the most useful metal in the fifth class. The members of this class melt with difficulty, and are not vaporized by the heat of an ordinary furnace. They are bivalent and form several oxides, sulphides, and chlorides.

1. *Iron.*—Pure metallic iron is of very rare occurrence in nature. The metal is found, however, in great abundance in combination with non-metals; its oxides and its sulphides being the most common forms. The *magnetic oxide* occurs in large quantities in many parts of the United States, and is used extensively in the manufacture of the metal. It is the ore from which the best "Swedish iron" is also made. In England, an impure carbonate, called *argillaceous iron ore*, is chiefly used. The iron of commerce occurs in three forms—cast-iron, wrought-iron, and steel.

I.—CAST-IRON.

A.—Cast-iron is a compound of iron with small and variable quantities of carbon. It is obtained from the ores by heating them in a blast furnace.

a. The blast furnace.—The construction of the blast

furnace may be understood from Fig. 31. The interior has the shape of a double cone. It is built of the most infusible firebrick, and inclosed in solid masonry. It is very large, perhaps fifty feet high by fifteen feet

Fig. 31.

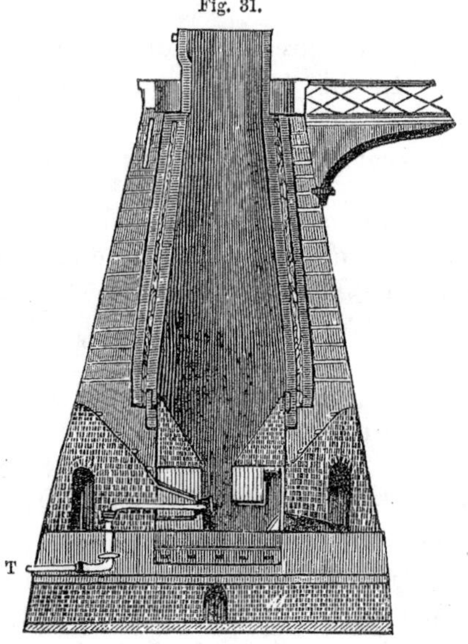

in width at its widest part. The bottom is closed, and the air needed for the fire is forced by steam-engines through pipes, T. The fuel and the ore, with broken limestone or other *flux* are put in at the top; the metal is drawn off at the bottom.

b. The process.—The ore, if necessary, is first roasted, by which it is changed to the form of *oxide.* The oxide mixed with fuel and flux *fills* the furnace. The fire having been started is kept up by the blast of hot air (a cold blast sometimes), driven by the engine. "Where the blast first touches the burning fuel, carbonic acid is formed; this gas, with nitrogen, rising through the furnace, comes in contact with white hot carbon, and is reduced to carbonic oxide. The layers of solid material thrown in at the top of the furnace gradually sink down, and as soon as a stratum of ore has gone far enough to be heated by the hot mixture of nitrogen and carbonic oxide, it becomes reduced to spongy metallic iron, which, mixed with flux and the earthy impurities of the ore, settles down to hotter parts of the furnace, where it enters into a fusible combination with carbon, while the flux and earthy impurities melt together to a liquid slag. The liquid carburetted iron settles to the very bottom of the furnace, whence it is drawn out at intervals through a *tapping-hole*, which is stopped with sand when not in use." (Eliot & Storer, p. 537.) The melted iron drawn from the tapping-hole is run into rough molds of sand, where it cools. This is the *cast* or *pig iron* of commerce.

II.—WROUGHT-IRON.

B.—Wrought-iron is nearly pure iron, but still contains a very small proportion of carbon. It is obtained generally from cast-iron by burning out its carbon in a reverberatory furnace.

a. The reverberatory furnace.—A section of a reverberatory furnace is shown in Fig. 32. The cast-iron is placed at D upon the hearth. The fire, A, is separated from the hearth by a wall of firebrick. The roof is arched downward to the chimney, which is forty or fifty feet high, to cause a strong draft.

b. The process.—Flame and hot gases from the fire striking against the arched roof, are reflected down upon the cast-iron. In a little time the iron begins to melt. When it has been all reduced to a pasty state the furnaceman unstops an opening (B), through which he puts his paddle. By thoroughly stirring (*pud-*

Fig. 82.

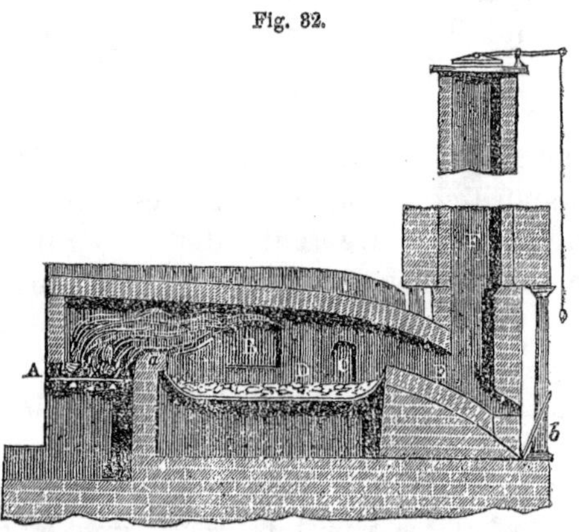

dling) the pasty mass, all parts are brought in contact with the hot air, during which a part of its impurities, in the form of slag or scoria, is allowed to run off, while its carbon is burned and its gas escapes to the chimney. The metal is then formed into balls; taken from the furnace; pressed or hammered to remove the remaining scoriæ; and then rolled into bars or other forms of *wrought* or *malleable* iron.

III.—STEEL.

C.—Steel is also a compound of iron and carbon, containing less carbon than cast-iron, but more than wrought-iron. It is made from cast-iron by burning out its carbon, or from wrought-iron by adding carbon to it.

a. From cast-iron.—From two to six tons of cast-iron

when melted is run into a large globular vessel, built of the most infusible substance. Numerous holes in the bottom of this crucible allow a strong blast of air to bubble up through the melted metal. A most violent combustion follows, the heat of which keeps the metal in a fluid state, while its carbon and a part of the metal itself are burned to oxides. Too much carbon is, by this process, removed, and a quantity of cast iron is added to restore carbon enough, to change the whole mass into steel. The crucible is then tipped upon its pivots, and the melted steel run off into molds. Less than half an hour is enough to change these tons of cast-iron into cast-steel. The process is of comparatively recent date, and is known as the *Bessemer* process.

b. From wrought-iron.—The older method of preparing steel is called "cementation." Bars of wrought-iron, imbedded in charcoal and inclosed in boxes from which air is very carefully excluded, are heated to redness, and kept in this condition for several days. A curious and obscure chemical action goes on, by which these two *solid* substances unite,—the carbon penetrating and combining with all parts of the iron, and thus changing it to steel. To make its composition uniform, this "blistered steel," as it is called, is melted and cast into ingots. The best quality of steel is made by this process.

2. *Illustrations of the class properties.*—All the other metals of the class more or less resemble iron both in physical and chemical properties. It needs the intense heat of the furnace to melt iron; the same is true of the others. Even at so high a temperature iron does not volatilize; neither do the others. Iron is generally bivalent, but sometimes quadrivalent; so are the rest,

with the possible exception of indium, which is only known to be bivalent.

(44.) Tin is the useful metal of the sixth class. The other members of the group resemble tin in their chemical properties. They are rare and unimportant.

1. *Tin.*—Tin is not an abundant element in nature, and yet it is one of the metals longest known to men. The mineral called *tinstone* (stannic oxide), is the chief source of the metal. Mixed with powdered coal and a little lime, the ore is spread upon the hearth of a reverberatory furnace. The carbon takes the oxygen from the ore, and the melted metal is run off into iron molds.

In color tin resembles silver. It is soft, malleable, and ductile.

Tin does not easily lose its luster by exposure to air, and on this account it is largely used as a covering for other metals: common tin-ware is made of sheet-iron, covered with a thin coating of tin.

(45.) The three metals of the seventh or tungsten class are all of rare occurrence, and of too little practical importance to be of interest to the general student.

(46.) The metals of the eighth or arsenic class in many respects resemble the non-metals of the trivalent group. They form the junction between the non-metals and the metals. Arsenic and antimony are interesting on account of the poisonous character of their compounds, and bismuth is used to some extent in alloys.

1. *Illustrations of the class properties.*—Arsenic has

been already described as a non-metal, and its relation to nitrogen and phosphorus pointed out. On the other hand it is related to antimony and bismuth whose metallic character is more decided.

Like arsenic, antimony combines with three atoms of hydrogen to form a combustible gas. Treated in Marsh's apparatus, it also forms a stain upon white porcelain, or a metallic mirror in the tube. So great is the resemblance between the stains of antimony and arsenic, that care is to be used not to confound the two metals in the test. The antimony stain is known by its more feeble luster, its blackness, and the high heat needed to volatilize it.

Bismuth forms oxides and chlorides whose composition is analogous to those of arsenic and antimony. It is trivalent like the others.

The metal itself is of a reddish hue and, like the other two, very brittle. It melts at 264° C. (507° F.), and, when cooling, it crystallizes and *expands*.

With other metals bismuth forms alloys remarkable for the low temperature at which they melt. Its alloy with lead and tin (2 parts Bi., 1 of Pb., and 1 of Sn.) is called "fusible metal:" it melts at 93° 7 C. (200° F.). This alloy is used for taking casts from medals and dies; on cooling, it expands, and, filling every crevice or line of the die, makes a most beautiful and faithful copy.

In bismuth the metallic character is very clear; in antimony less evident; in arsenic quite doubtful; in phosphorus hardly to be found; and in nitrogen, absent. In this group the metals and the non-metals form a junction; from nitrogen to bismuth the transition is gradual and perfect.

(47.) Lead and thallium, the metals of the ninth class, are alike in many of their physical properties. Lead is obtained from galena, which is its most abundant ore, and is used for many familiar and important purposes.

1. *Lead and thallium.*—Metallic lead is too common to need a lengthy notice of its physical properties: thallium, on the other hand, is known to few, having been discovered in 1861; but the description of one will almost answer for that of the other. Both are bluish-white when freshly cut, but their surfaces soon tarnish when exposed to air. They are alike soft enough to yield to the finger nail: both are fusible below a red-heat, malleable and ductile.

In their chemical properties these metals do not so perfectly agree. Lead is bivalent: thallium is univalent. In other respects, also, thallium resembles the alkaline metals rather than lead.

2. *Lead obtained from galena.*—The ore from which the lead of commerce is obtained is a sulphide (Pb S), called galena. The color and luster of this ore is much like that of the metal itself, but the ore is much harder. It is crystalline, and sometimes occurs in cubes of the most perfect form.

To obtain metallic lead, galena is mixed with lime and roasted in a reverberatory furnace. By this means a part of the ore is changed to oxide, another part to sulphate, and some remains as it was. The air is then shut off from the furnace and the heat raised: the compounds are all decomposed, and metallic lead is produced.

3. *Its uses.*—Metallic lead has numerous and important uses. Among them we may especially notice

the construction of water pipes and cisterns. In cities supplied with water from reservoirs, the conduit pipes are almost universally made of lead.

But since solutions of lead are poisonous, the chemical action of lead upon water is a matter of great importance. Together, especially in the presence of air, they form an oxide which is somewhat soluble in water. The corrosive action is very much modified by the presence of impurities. It is increased by nitrates and chlorides: it is diminished by sulphates and carbonates.

Water, containing a solution of calcic carbonate, scarcely affects the lead; but if it contains much free carbonic acid, this removes the lead carbonate, which otherwise would protect the surface, and leaves the metal constantly exposed to corrosion. "It has been proved by numberless experiments that the action of natural waters upon lead is so general that it is rare to find any sample of water, which has been kept in a leaden cistern, wholly free from traces of that metal. The opinion of most chemists is at this time (1867) decidedly adverse to the use of leaden pipes in houses, in spite of the fact that the metal is nowadays employed for this purpose almost everywhere with apparent impunity." (Eliot & Storer, p. 493.)

(48.) Copper, mercury, and silver are found native, but more generally as sulphides. They do not decompose water at any temperature. Nitric acid speedily changes them to nitrates, at the same time evolving nitric oxide. Copper is not readily acted upon by air; mercury only at a high temperature, and silver not at all.

I.—COPPER.

A.—Copper is widely distributed in nature. It is a red metal, very tenacious and ductile. Brass is an alloy of copper and zinc.

1. *Copper is widely distributed.*—In small quantities copper is found in the soil almost everywhere, and traces of it exist in plants and animals. It occurs native in many places, but in far greater abundance it is found in combination. The sulphides are the most common ores from which the metal is obtained.

2. *Its properties.*—Sheet copper and copper wires are familiar forms of the metal in commerce, and they suggest its malleability and ductility. These properties, together with its great tenacity, render it valuable for many purposes in the arts. It slowly tarnishes in the air, and is readily acted upon by vegetable acids: the compounds formed are, many of them, poisonous. On this account copper vessels to be used for culinary purposes are usually coated with tin. Copper is among the very best conductors of electricity, and is much used in the construction of electrical apparatus and lightning-rods.

3. *Its alloys.*—Several alloys of copper are of great importance. Brass, bronze, and gun-metal are examples. The first is an alloy of copper and zinc, the others of copper and tin. German silver is another alloy containing copper; its other constituents are nickel and zinc.

II.—MERCURY.

B.—Mercury is found native, but the chief source of

the metal is cinnabar (a sulphide), found in considerable abundance in Spain and California. It is the only metal which at common temperatures is a liquid: its melting point is $-39^\circ 6$ C. ($-39^\circ 4$ F.).

Many of its uses are already familiar to the student. Among others we notice that it is of great use in extracting gold and silver from their ores; in silvering mirrors; and that some of its compounds are highly prized in medicine. Calomel is the mercurous chloride ($Hg\,Cl$).

Another chloride, the mercuric chloride ($Hg\,Cl_2$), is the well-known *corrosive sublimate*, a virulent poison. Alloys of mercury are called *amalgams*.

III.—SILVER.

C.—Silver is sometimes found native, but generally in combination; the sulphide being the most important ore. It occurs in small quantities in galena, from which it is obtained by cupellation: from other ores it is obtained by amalgamation. It does not tarnish in pure air, and its alloys are largely used for coin and for "silver ware."

1. *Silver in nature.*—Silver, like the other metals of greatest use in the arts, is very widely distributed. It is found native and often alloyed with mercury, copper, and gold, but its sulphide ($Ag_2\,S$), either alone or mixed with other metallic sulphides, is its most common form. From these ores the metal is usually obtained. They are found in many countries of Europe, but in greater abundance and richness in Peru, Mexico, and the Pacific slope of the United States.

2. *Cupellation.*—Galena almost always contains silver, and often in quantities which make it valuable for the extraction of silver as well as lead. "It has been found profitable to extract the silver from lead that contains less than one thousandth of its weight of the precious metal." From lead rich in silver the precious metal is obtained at once by *cupellation;* but when so poor as that just described, the lead is first melted and then, while cooling slowly, it is stirred. As it cools it crystallizes, and the crystals may be dipped out of the liquid mass in colanders. Now an alloy of lead and silver will remain melted when cooled below the temperature at which pure lead solidifies. Hence the crystals dipped out in the colander are crystals of *lead*, while *all* the silver remains in the fluid left behind. By this means the proportions of lead may be reduced until the metal is rich in silver, after which it is *cupelled.*

The process of *cupellation* is based upon the fact that lead is rapidly oxidized by air at high temperature, while silver is not.

The alloy, placed in a shallow, porous vessel of bone-earth, called a cupel, is melted in a furnace, and its surface is at the same time exposed to a current of hot air. The lead is changed to an oxide; the melted oxide is partly absorbed by the cupel, while another part runs off into other vessels. The silver is not affected by the air, and when the lead has passed away, the precious metal still remains in the cupel.

3. *Amalgamation.*—From other ores than argentiferous lead, silver is extracted by a process called *amalgamation.* It is based upon the fact that silver is very soluble in mercury.

For this process also, the ores require a preliminary

treatment. After being thoroughly ground they are mixed with common salt, and then roasted. By this means the silver is changed to a chloride. The roasted substance is then mixed with water and fragments of iron, and the mixture is violently shaken in revolving casks. Mercury is soon added and the agitation is kept up for, perhaps, twenty hours. In the mean time the iron decomposes the chloride, and the silver thus set free, is dissolved at once in the mercury. The excess of mercury in the amalgam is removed by filtering it through leather bags under pressure: and the rest by distillation. The precious metal is left behind.

4. *Its alloys.*—Silver is too soft for most purposes in the arts. Small quantities of other metals increase its hardness. For coin and for articles of plate it is alloyed with copper. The English *standard silver* contains 7.5 per cent. of copper. The standard is regulated by law, and is not the same in all countries. In the United States the legal silver coin is $\frac{9}{10}$ silver and $\frac{1}{10}$ copper.

(49.) The metals of the gold class are always found in the metallic state. They are not tarnished by air at any temperature short of their melting points, nor can they be dissolved by nitric acid. Only chlorine or aqua regia can dissolve them. Gold and platinum are the most familiar and important.

1. *Gold.*—In small quantities gold is very widely distributed in nature. Fine grains of it occur in the sands at the bottom of many rivers; in the crystalline rocks and the soils derived from them. Iron pyrites, found almost everywhere, often contains traces of gold; and silver is never found entirely free from it. Few

places however seem to possess the precious metal in quantities that will pay for the laborious work of separating it from the sands or rocks in which it is found.

Except iridium and platinum, gold is the heaviest of metals. It is very much heavier than sand, so that when gold-bearing sand is violently shaken in water, the fine and precious grains quickly sink to the bottom, while the sand may be poured off with the water. By repeating this process, called "washing" the sand or other loose material is finally all washed away. The fine metallic grains left behind are seldom pure gold. The baser metals—silver, copper, and others—are taken out by sulphuric or nitric acid in which they will dissolve, but which can not dissolve the gold. This process of separating gold from the metals with which it is alloyed is called *refining*.

When the native gold is scattered in fine grains through solid rock it is extracted by *amalgamation*. Mercury mixed with the crushed ore dissolves the gold, and it is then taken from it by filtration and distillation.

2. *Platinum.*—Platinum is a rare metal, not so widely distributed as gold. The slopes of the Ural mountains, Brazil, and Peru are among the localities richest in this metal. It is always found in the metallic state, but never pure. It is heavier than gold, and is obtained from loose sands or soils in the same way. Its commercial value is about one-half that of gold; about eight times that of silver.

Pure platinum is almost as white as silver, and can not be tarnished in air at any temperature. The most intense heat of the blow-pipe is needed to melt it. On these accounts platinum vessels are much used in the laboratory. Few, indeed, are the chemicals which can affect it. Aqua regia and the caustic alkalies are among those which can. On the other hand it readily forms alloys with most other metals, and these are more easily melted and are soluble in acids.

CHAPTER IV.

ON DECOMPOSITION IN PRESENCE OF AIR.

General Statement.—Organic substances, or hydrocarbons obtained from them, when exposed to the action of air under the influence of a proper amount of heat or moisture, or both, may be decomposed. Carbonic dioxide and water are the chief products of the action.

I.—Combustion.

(50.) Combustion is a mutual chemical action, generally between oxygen and some other substance, and which, when rapid enough, evolves heat and light. A certain temperature must be reached before a substance will kindle, but once started the chemical action will produce heat enough to keep the action going.

Fig. 33.

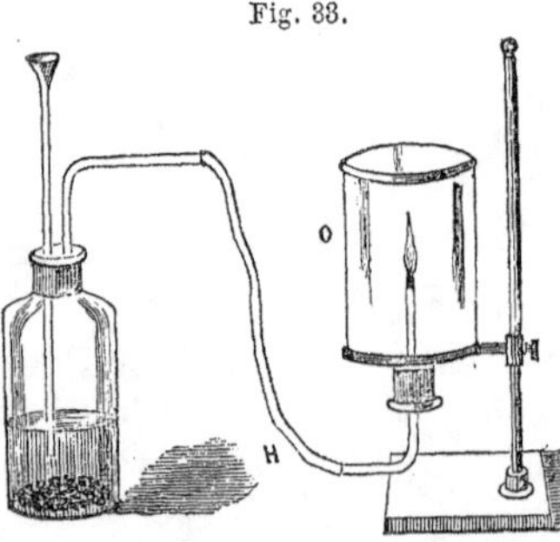

1. *Combustion a chemical action.*—Combustion is among the most familiar phenomena of every-day life. It is a mutual chemical action between at least two substances. To illustrate, let us refer to the familiar case of oxygen

and hydrogen. Fig. 33 represents a jar of oxygen, with a jet of hydrogen burning in it. It is the hydrogen which appears to burn. In Fig. 34 oxygen is flowing from a jet and burning in a vessel kept full of hydrogen—in this case the oxygen appears to burn. The truth is, that in both cases alike the two gases are combining to form the liquid water.

Fig. 34.

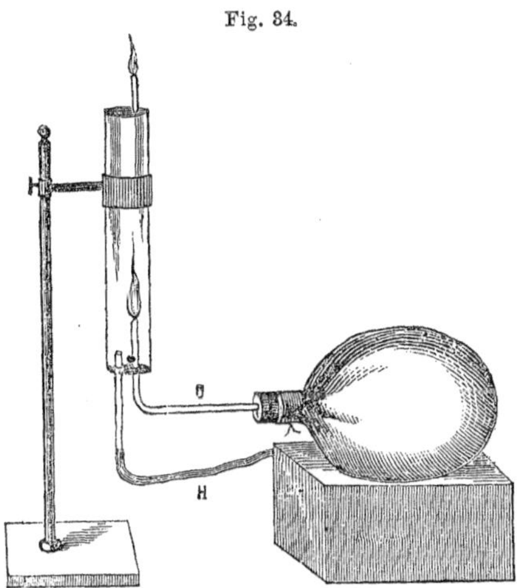

2. *Between oxygen and other substances.*—In all ordinary combustion, oxygen is one of the active substances. When wood burns in air, it is because the carbon and hydrogen of the wood combine with the oxygen of the air. Few, indeed, are the exceptions to this—the burning of a wax taper in a jar of chlorine (p. 112) is one.

3. *When rapid it evolves light and heat.*—The evolution of light and heat accompanies all familiar cases of burning, but the same chemical action going on

more slowly, gives off heat without light. An iron wire, when burned in pure oxygen, produces a splendid light, but when allowed to slowly rust in air, no light is ever seen. The chemical action is the same in both cases—oxygen combines with the iron and they form an oxide. Moreover, to change the iron to an oxide takes the same amount of oxygen in the two cases, and still further, the same aggregate quantity of heat is given off. Both are cases of combustion, the one being rapid, the other slow.

The *amount* of heat depends upon the amount of material, especially of oxygen taking part in the action; the *intensity* depends upon the rapidity with which the action goes on.

4. *A certain temperature is needed.*—Thin shavings of the driest pine wood, as is well known, will rest in air for any length of time unburned, but when heated by a burning match, how quickly do they disappear! The match, by its heat, simply raises the temperature of the wood until having reached a certain point, the oxygen of the air begins rapidly to combine with its elements. The temperature at which a substance begins to burn is tolerably constant, and it may be called its *kindling point.* The kindling point varies greatly in different substances—phosphorus, for example, inflames, sometimes, by the gentle heat of the hand, while sulphur must be heated to about 560° before it will take fire, and ordinary fuels to about 1,000°.

5. *Temperature kept up by chemical action.*—But let the burning once begin, and the kindling point is kept up by the combustion itself—a match may kindle the fire which destroys a city, but if by any means the burning body can be cooled below the kindling point,

the fire is quenched. Over the flame of a gas-jet, press down a sheet of wire gauze (Fig. 35); the gas goes through the gauze, the flame does not. The cold metal reduces the heat, cooling the gas below its kindling point. The metal finally becomes red-hot, after which the flame freely passes it. Upon this principle the "safety lamp" is made (Fig. 36). It consists of a wire gauze cylinder completely enveloping the burner of the lamp. The combustible fire-damp of a mine may burn some time inside the cylinder, *warning* the miners of its dangerous presence before the cold meshes of the gauze will allow the flame to pass out and explode the mine.

Fig. 35.

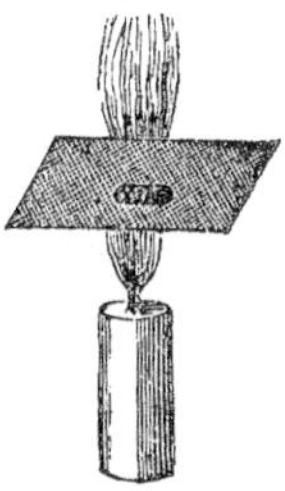

(51.) The substances used for fuel in all ordinary cases are compounds of carbon and hydrogen. The products of combustion are water and carbonic dioxide. To burn with flame, the body must be either a gas itself, or must give off a gas on being heated. In common flames we may notice three parts—the nucleus, the luminous envelope, and the non-luminous envelope.

Fig. 36.

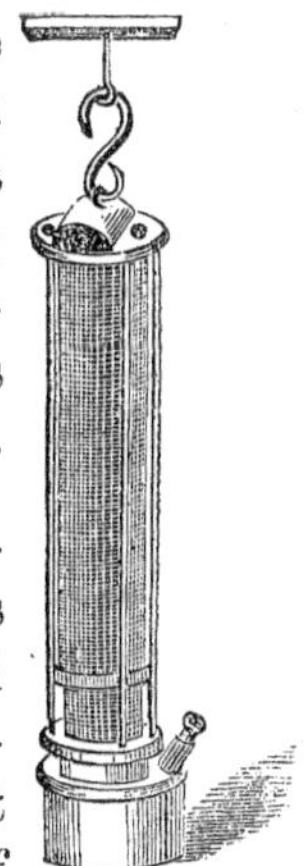

1. *Fuel.*—All fuel is of vegetable origin. That wood and charcoal are so is evident—not less so are all forms of coal found in the earth. They are the remains of an ancient vegetable or forest growth which, under the influence of heat, pressure, and chemical force, have at last parted with their oxygen and hydrogen, while the carbon still

remains. The resins, oils like petroleum, and coal gas, more rarely used for fuel, are substances, also, which have come from plant bodies. But all these substances are compounds of carbon and hydrogen in different proportions.

2. *The products of combustion.*—Upon the bottom of a glass jar stand a burning wax taper, and cover the jar tightly. The flame soon grows dim and is finally extinguished. Let the taper be removed and a small quantity of lime-water poured into the jar. The milkiness of the fluid shows the presence of *carbonic dioxide.* The carbon of the wax has united with the oxygen of the air.

Again, over the flame of a burning gas-jet hold a cold glass jar. The sides of the jar are quickly dimmed with dew, which must have been formed by the hydrogen of the wax uniting with the oxygen of the air.

Water and carbonic dioxide are the important products of all ordinary combustions.

3. *Gases only burn with flame.*—The flames of all ordinary combustion are due to burning gases. The flame of a candle is as truly a gas flame, as is the flame of a jet of illuminating gas. The solid wax or tallow in the wick must be first melted and then vaporized by the heat of the match, before it will take fire. The gas once lighted causes a flame, whose heat melts the wax below, and the liquid, lifted by capillary force, through the wick, is changed to vapor as fast as it is needed in the flame. Take the wax in a test-tube with a jet pipe through its tight cork, and heat it. The wax melts—boils—the vapor at length issues from the jet, and when touched with a lighted match burns (Fig. 37) with exactly the same kind of flame, only less steady.

Alcohol and oil burn with flame only after heat has changed them to vapors. Bituminous coal and wood, burn with flame because they contain substances which the heat applied can vaporize. While hard coal burns with no flame, because it gives off no gases when heated.

Fig. 37.

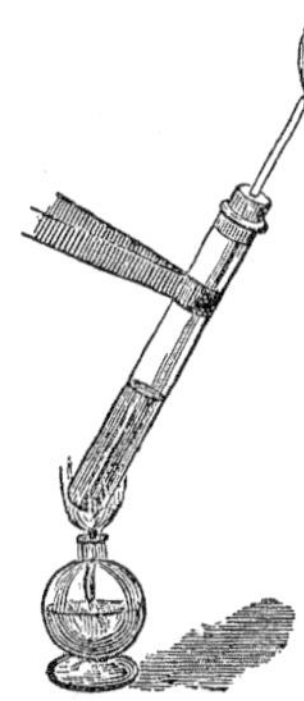

4. *Three parts to a common flame.*—The dark center and the luminous cone of common flames have been noticed by us all. This dark center or nucleus consists of the gases formed from the fuel; but they are not in a burning condition. The luminous envelope consists of these gases combining with the oxygen of the air. In this part of the flame only does combustion occur. Outside of this is an envelope made up of water-vapor and carbonic dioxide produced by the burning.

The *nucleus is not burning*, for the end of a match, plunged to the center of the flame, does not burn while there: it takes fire while coming out. And even gunpowder may rest in this center of a flame unharmed! Invert a dinner plate or common bowl upon the table; its bottom is a very shallow dish which will hold a small quantity of alcohol. Put some gunpowder on the end of a small cork and place it at the middle of the plate and then touch the alco-

Fig. 38.

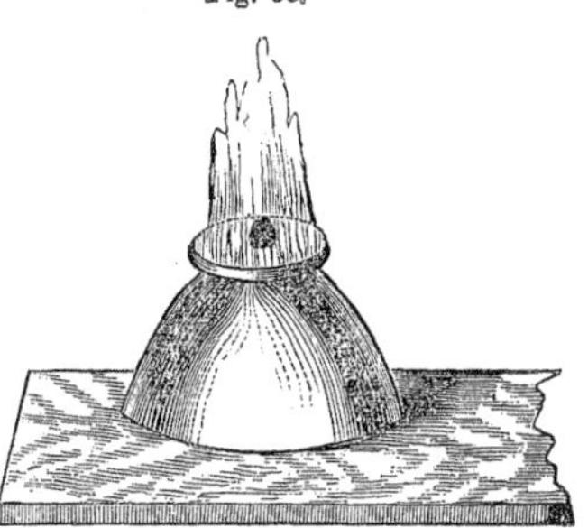

hol with a match. A large flame is formed, in the center of which the powder remains unharmed (Fig. 38), until some air current wafts the flame or the alcohol is nearly consumed.

The *burning takes place in the luminous envelope.* Over the flame of an alcohol lamp suddenly lower a sheet of writing paper, holding it for a moment across the middle of the flame. Remove it, and a scorched ring will be seen; the paper is burned just where it was in contact with the luminous envelope.

Hold a cold glass plate in place of the sheet of paper for a moment on the flame: a ring of dew is condensed upon it where it comes in contact with the non-luminous envelope.

(52.) Combustion is resorted to merely for the heat and the light it produces, not as in other chemical processes, for the material which it forms.

I.—HEAT.

A.—To give the greatest heat the air should be thoroughly mixed with the burning gas, as in the Bunsen's burner, a common furnace and the oxyhydrogen blow-pipe.

Fig 39.

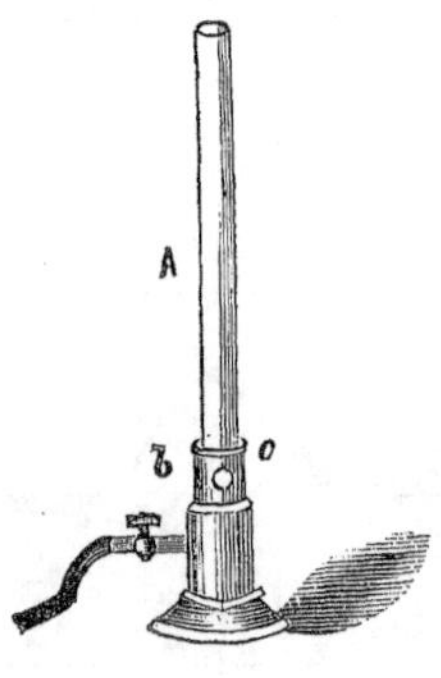

1. *The Bunsen's Burner.*—(Fig. 39) represents a Bunsen's burner. The gas is brought from the chandelier by means of the rubber tube, and issues from a jet pipe inside the tube A. Just below the end of this jet are two large holes on

opposite sides of the tube, one of which is shown at *b*. A movable collar *c*, with corresponding holes, may be turned around the tube so as to open and close these holes at pleasure. Now a constant supply of *gas* from the jet and of *air* entering at the holes keeps the tube (A) full of a *mixture*, which is lighted as it issues from the upper end. This burning mixture gives an intense heat; a glass tube is quickly softened or melted by it; but the light is very feeble.

2. *The furnace.*—In the common stove and furnace, the air entering at the draught *mixes* very thoroughly with the fuel, and the heat is much more intense than it would be if the two came in contact only at the surface.

3. *The oxyhydrogen blow-pipe.*—A mixture of hydrogen and oxygen, in the right proportions to form water, burns with heat of surprising intensity. By means of the oxyhydrogen, or compound blow-pipe, this flame may be obtained with little danger of explosion. Notice Fig. 40, which shows a section of the jet, and Fig. 41, which shows the instrument in use.

Fig. 40.

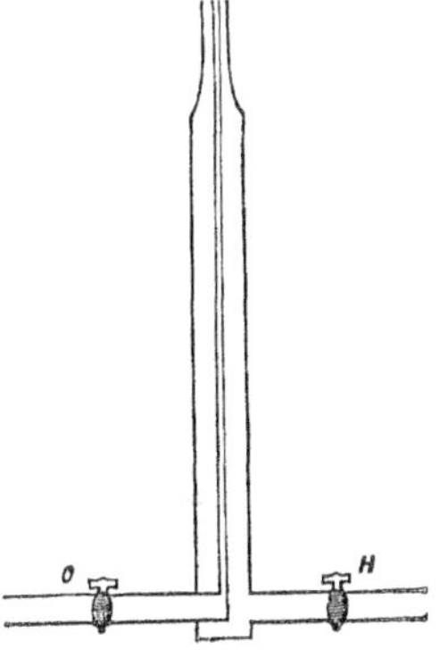

The gases are stored in separate bags. Hydrogen passes from one of these into the larger or outside tube of the jet, and flows out at *c*, while oxygen from the other bag passes through the inside tube, and out at the same point. It will be seen that the two gases *mix* just at the point of the jet, and that here the combustion takes place.

The heating effects of this flame are astonishing. Zinc and antimony are vaporized by it. Iron and steel (Fig. 41) burn in it like thread in a lamp flame. Platinum, and even quartz and other rocky matter,

Fig. 41.

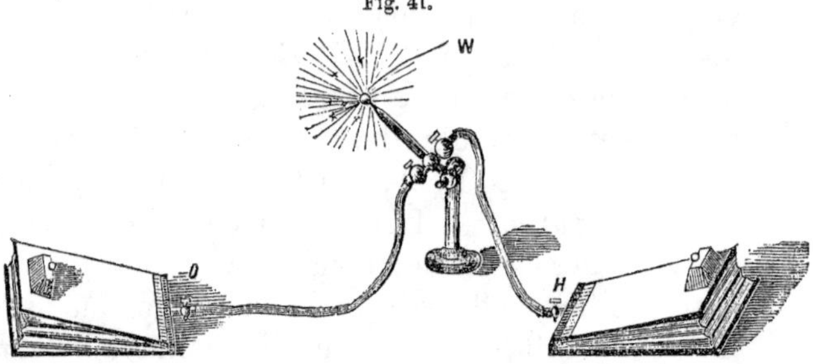

may be melted or softened. Only the electric heat can surpass it.

In all these heat flames the light is feeble.

II.—LIGHT.

B.—To give the greatest light there should be a full supply of air to the surface of the gas-jet, as in the common lamp, the argand burner, and ordinary gas flames.

Fig. 42.

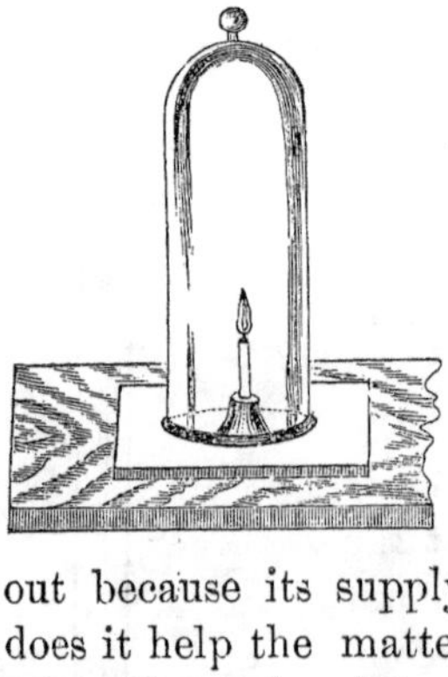

1. *A full supply of air.*—Witness the following experiments. Let a glass jar be inverted over a burning taper (Fig. 42). The flame continues for a time, then dies. It goes out because its supply of oxygen is exhausted. Nor does it help the matter much to raise the jar a little space above the plate on which the taper stands, for

the jar remains full of impure air, which keeps the pure air out.

Now again, put the taper at the bottom of a jar whose open mouth is upward, and, if necessary, partly cover it (Fig. 43). The flame dies almost as quickly as before, and for the same reason. We learn from these experiments that a supply of fresh air is absolutely needed in combustion.

Fig. 43.

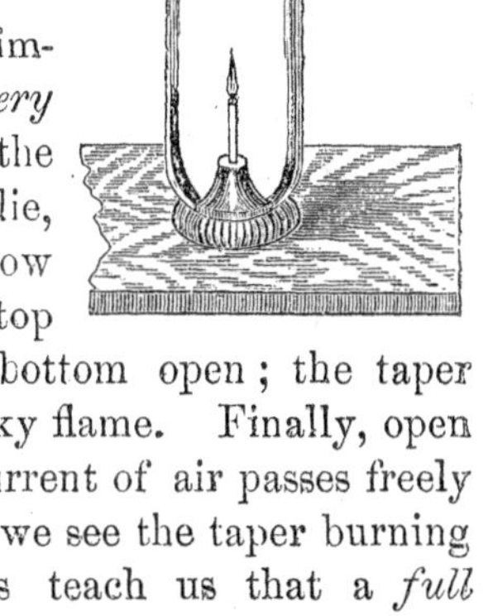

Go further—take a lamp chimney and so place it that a *very small* opening admits air at the bottom—the flame does not die, but burns dimly. And now again: almost cover the top of the chimney leaving the bottom open; the taper burns, but with a dim or smoky flame. Finally, open both top and bottom. A current of air passes freely up through the chimney, and we see the taper burning brightly. These experiments teach us that a *full supply* of air is necessary to produce a luminous flame.

But we have seen that when a full supply of air is *mixed* with the burning gas an almost non-luminous flame is produced. A full supply of air must be brought in contact with the *surface* of the gas-jet if the greatest light is to be obtained.

2. *This is done in the common lamp.*—The oil or kerosene lamp is too familiar to need description. Notice that the *flat* wick usually employed exposes a large surface to the air, and that the chimney, open at the top and bottom, allows a constant current of fresh air to pass up through it. By this means an abundance

of air is brought in contact with the flame, at its surface only, and the bright light is the result.

3. *The argand burner.*—In the argand burner an artifice is resorted to, by which the surface of the flame exposed to air is increased. The wick is a hollow cylinder, and the air passes up through the inside of it as well as around its outside. The gases from the wick expose a double surface to the air and give off a greater light accordingly.

4. *The gas burner.*—The burner of a gas chandelier is so made that the gas escapes in a fan-shaped jet. This is done in many ways. In the end of some burners we may notice two small holes, and by putting pins into these we find them to be the ends of two little tubes slanting toward each other, so that if continued outward they would meet. Now the two jets of gas from these tubes strike against each other with force enough to flatten both out into a single fan shaped jet. In this way a large surface of gas is exposed to air without making a mixture of the two substances, and the luminous flame is produced.

(53.) The light of a flame is due to solid particles, or dense vapors, intensely heated.

1. *To solid particles.*—The luminous power of solid bodies in a heat-flame is illustrated by the so-called Drummond or calcium light. It is made by simply placing a little ball of lime in the flame of the oxyhydrogen blow-pipe. So very intense is this light that the eye is blinded by its direct rays; and, used in lighthouses, it has been seen miles away at sea.

Observe now; lime remains solid even in the most intense heat of the blow-pipe, but the burning mixture gives heat enough to raise the lime to a white heat and in this condition it shines with a blinding light.

Doubtless on this principle we may, in part, account for the light of common flames. We know that the burning gas is being decomposed and that one of its constituents is carbon. We know, too, that carbon is a solid, even at the highest heat, but that heated in air it combines with oxygen. At the instant, then, when set free from the burning gas, and before its union with oxygen, each little molecule of carbon is heated white-hot, and, shining brightly, adds its mite to the light of the flame.

2. *To dense vapors.*—But that all the light of flames is not to be explained in this way is seen clearly from the fact that in some of the most dazzling there is no substance present which can remain solid at the temperature of the flame. Remember the blinding light of phosphorus burning in oxygen gas. "Now phosphoric anhydride, the product of this combustion, is volatile at a red-heat, and it is, therefore, manifestly impossible that this substance should exist in the solid form at the temperature of the phosphorous flame, which far transcends the melting point of platinum."* (Dr. Frankland.) Many other examples of a similar kind might be given. The light of such flames, and doubtless much of the light of all flames, is due to intensely heated dense vapors.

II.—Respiration.

(54.) Respiration is a chemical action similar to combustion. Large quantities of air are, by it, made

* See Chemical News, American reprint, vol. iii., p. 237.

unfit to support life, hence the supreme importance of ventilation.

1. *Respiration a chemical action.*—The walls of the air cells of the lungs are covered with a net-work of minute capillary blood-vessels. Into the air cells successive portions of fresh air enter, to be at once thrown out again, while, at the same time, the impure blood of the system is constantly coursing through these vessels. What changes are being made, first in the air, and second, in the blood?

If one breathes through a tube into a vessel of lime water, its milky color soon shows the presence of *carbonic dioxide.* If he breathe on a cold surface, the moisture condensed shows the presence of *water vapor.* If these substances be removed from the breath which has been collected in a jar, the flame of a taper, afterward inserted, is extinguished, showing the absence of *oxygen.* The gas that remains is chiefly nitrogen. Notice: the air in the lungs gives up its oxygen and receives carbonic dioxide and water vapor.

Could we examine the blood, we should find that while in the lungs, its color changes from purple to bright red, due to the loss of carbonic dioxide and water vapor and the receipt of oxygen.

Now how can these changes be explained? It has been found that particles of the body itself are constantly being worn out. Not an act can be done, a word spoken, nor can a thought occur without the disintegration of some portion of the organs. These waste particles are in great part thrown into the blood: they are its impurities. These particles are composed chiefly of carbon and hydrogen. The pure blood is charged with oxygen, and as it flows, this gas combines with the carbon and hydrogen of the waste, at all points in its course, and the carbonic dioxide and water vapor, thus formed, pass into the lung cells to be finally exhaled from the system. The action is a process of slow combustion. The waste particles are the fuel, oxygen is supplied, and carbonic dioxide and water vapor are the products.

2. *Large quantities of air spoiled.*—It will be at once seen that the air of our rooms is being constantly made unfit to keep up this action by which the blood is purified. In the first place its oxygen is being taken out, and in the second place impurities are being thrown into it. Not only does it receive the worn-out matter of the system from the breath, but we may here add that other portions of waste, even more offensive, are incessantly being given into it from the body by perspiration. The total amount of impurity thus added to the air of a close room is frightful, since, upon the average, about 20,000 cubic inches of air pass through the lungs of a single person every hour!

3. *Hence the need of ventilation.*—As the flame of a taper dies in a closed jar, so a human being would die if confined long enough in an air-tight room. As the flame flickers and burns dim when a small supply of air is furnished, or the products of combustion are not removed, so the life of human beings flickers and grows feeble in rooms where fresh air is not supplied

III.—Decay.

(55.) The decay of wood or other vegetable matter is a slow process of combustion. By the gradual loss of carbonic dioxide and water, the decaying wood is finally changed into a brown or black mold to which the term humus is given.

1. *The decay of wood.*—If fine sawdust is thoroughly moistened and put into a stoppered bottle containing air, and afterward exposed to a temperature of about 16° C. (60° F), it will after a long time be found partially decayed. By the usual tests it will be found that the air in the bottle has given up a part of its oxygen and received carbonic dioxide in return.

Wood of whatever kind, when exposed to the continued action of moist air and warmth will, like the sawdust, in the experiment, be gradually decomposed. Moisture and warmth are essential to decay, since it is found that wood exposed to the cold of the Arctic regions, or to the dry air of Egypt, will remain, even for centuries, in good condition.

2. *Other vegetable matter.*—If a flask (Fig. 44) partly filled with peas and water is furnished with a bent tube reaching over to an inverted larger tube filled with water, there may after a time be seen bubbles of gas rising into the tube. This gas, if tested, will be found to be carbonic dioxide, and when the peas are examined, they will be found to be partially decayed.

Fig. 44.

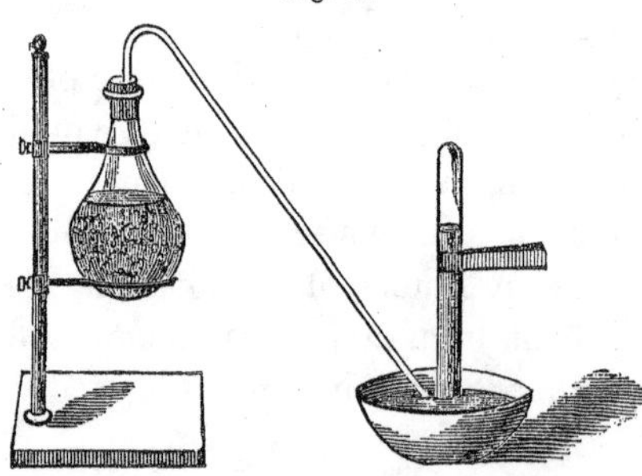

3. *A slow combustion.*—Now the experiments with

the sawdust and the peas are simple illustrations of what occurs whenever any kind of vegetable matter is long exposed to the action of air under the influence of moisture and warmth. They are slowly decomposed; a part of their carbon and of their hydrogen unites with oxygen to form carbonic dioxide and water, while the rest of these elements in combination with oxygen remain in the form of a loose, solid, black mold.

The chemical action in the process of decay is, clearly, very similar to that of combustion. Indeed it differs from combustion chiefly in being less rapid. Heat is evolved in decay as in combustion, and from equal quantities of material the same amount. In some rare cases a feeble light is also given off by decaying vegetable matter. Decay is to be considered as but a slow process of combustion.

4. *Humus.*—The brown or black mold that is left after the decay of plants, is called *humus*. This term, however, is not the name of a single substance, but rather of a mixture of several compounds of carbon, hydrogen, and oxygen in various proportions. Humus gives to fertile soils their rich brown or black appearance.

CHAPTER V.

ON DECOMPOSITION IN ABSENCE OF AIR.

General Statement.—Organic substances from which air is wholly or in part excluded may be decomposed by the action of heat or moisture. The products of the action depend upon the substance acted on, and the circumstances under which the action occurs.

I.—DESTRUCTIVE DISTILLATION.

(56.) When wood or other vegetable substance, is heated in close vessels, it is decomposed The process is called destructive distillation. Charcoal, a mixture of gases, pyroligneous acid, and wood-tar, are the products. From the last two products several other remarkable substances may be obtained, among which we notice methylic alcohol, creosote, and paraffine.

1. *Destructive distillation of wood.*—Let the following experiment be tried. Into a small flask (Fig. 45) put some fine splinters of some hard wood, beech wood answers the purpose well. Let the flask be tightly corked, and provided with a bent tube reaching over into a bottle which is kept cold by the water which surrounds it. On heating the flask the wood soon turns black; volatile matter is driven over, some of it being con-

densed in the bottle, while another part escapes in the form of gas. This gas may be collected in a second bottle over water, by passing it through a rubber tube reaching from the short tube *t*. The black solid left in the flask is charcoal; the gas collected is a mixture of several,—marsh gas and olefiant gas being among them, while the fluid in the first bottle consists of pyroligneous acid and wood-tar.

Charcoal and the gases have already been described.

Fig. 45.

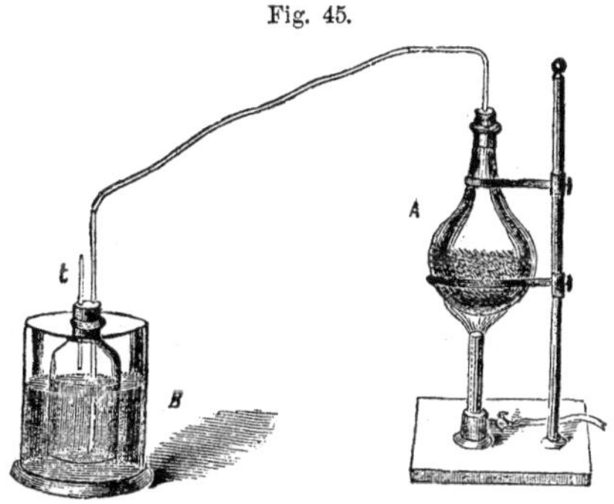

2. *Pyroligneous acid.*—Pyroligneous acid is often called *wood-vinegar*. Dry beech wood yields it in greatest abundance. It is a dark brown liquid, with a sour, smoky taste. It contains acetic acid, and on this account has been largely used in making such acetates as are employed in the arts, especially in calico printing and dyeing. Sodic and plumbic acetates are examples.

3. *Wood-tar.*—Wood-tar is a very dark-colored resinous fluid. There are several varieties,—one, largely used in ship-building and other arts, is obtained by a rude distillation of resinous pine wood; another is obtained from hard wood. It is sometimes used as a

covering for wood to preserve it from decay: the more volatile constituents of the tar passing away, leave the harder part (pitch) in the pores of the wood. Water is thus kept out of the pores of the wood, and this, together with the action of creosote, to be soon noticed, prevents decay.

4. *Methylic alcohol.*—When pyroligneous acid is distilled, a very volatile liquid may be obtained, which, being afterward rectified by the use of quicklime, constitutes the *methylic alcohol* of commerce: it is often called *wood spirit.* It is a limpid liquid, very inflammable. It may be used instead of alcohol for many purposes, especially for dissolving resins in making varnish.

In its composition methylic alcohol may be described as water in which one atom of hydrogen has been replaced by the radical $C\,H_3$ (methyl). The formula for water being written $H\,H\,O$, that of the alcohol ($C\,H_4\,O$) may be written $C\,H_3\,H\,O$.

5. *Creosote.*—The smoky taste of pyroligneous acid and the power of both this acid and wood-tar to prevent decay, is due to the presence of a curious compound of carbon, hydrogen, and oxygen ($C_8\,H_{10}\,O$) called *creosote.* One pound of the acid contains about a quarter of an ounce of it in solution. Its most curious and valuable property is its power to prevent decay: indeed it is the most powerful antiseptic known. Flesh remaining a few hours in a fluid made by dissolving 1 part of creosote in 100 parts of water, will not afterward decay. That meats are often *cured* by exposure to smoke, is familiar to all: now the preservation and the peculiar flavor of smoked meat is due to the action of creosote in the smoke. This substance is often used in medicine, but

when taken internally, except in very small quantities, it is a corrosive poison.

6. *Paraffine.*—Paraffine may be obtained by distilling wood-tar. It is a crystalline solid with neither taste, color, nor smell. Its most remarkable property is its indifference to the chemical action of other substances. It can resist the action of the strongest alkalies and of the most corrosive acids. It is, however, combustible, and its flame is white and smokeless. Paraffine candles rival the most costly wax candles in luster and in the strength and beauty of their light.

7. *Other substances.*—Numerous substances, besides the few just described, may be obtained by the destructive distillation of wood. Each different kind of wood and of other vegetable matter yields some different products, but for the most part they are all compounds of carbon with either hydrogen or oxygen, often with both.

It should be noticed that none of these products are supposed to exist ready formed in the plant. By heat the substance of the plant is broken up; its elements are rearranged and these hydrocarbons formed.

(57.) By the destructive distillation of bituminous coal, and sometimes of other substances illuminating gas is made. The material being heated in iron retorts, its volatile constituents are driven off. The gas thus formed is purified by passing first through cold pipes, and then over lime. It is afterward collected in gas-holders, from which it is pressed out into the pipes of the city and through these into the chandeliers of the houses.

1. *Bituminous coal.*—In the United States alone, there are about 130,000 square miles of workable coal-

fields. The coal found so abundantly in nature is called *mineral coal;* it consists of carbon mixed with other matter, especially with volatile compounds of carbon and hydrogen. The two varieties of mineral coal—*anthracite* and *bituminous*, differ chiefly in the amount of their volatile constituents. The first contains very little of the hydrocarbons, is very hard and burns with a very feeble bluish flame; the second contains a large amount of volatile compounds, is much softer, and burns with a bright flame. From bituminous coal illuminating gas is generally made.

2. *Other material.*—Other material is sometimes used. Wood, when heated in close vessels yields an excellent gas. Resins furnish gas of superior quality, and are often used in its manufacture. Crude or refuse oil, unfit for burning, is sometimes used. But on the score of economy none of these substances can compete with coal.

3. *Heated in iron retorts.*—The vessels in which the coal is heated are called retorts. They are usually of iron, about seven feet long, and scarcely more than a foot in diameter. Fig. 46 shows the ends of five of these retorts placed in a single furnace. Each one, after receiving a charge of from 100 lbs. to 150 lbs., is closed air-tight as shown at *G*, and made red-hot by the fire in the furnace, whose doors are shown at A. In the course of a few hours the volatile matter of the coal is driven off: the residue called coke is then raked out, cooled, and used for fuel.

4. *Its volatile constituents.*—The gaseous mixture driven off by heat, contains olefiant gas, marsh gas, carbonic dioxide, hydrogen, ammonia, hydro-sulphuric acid, and coal-tar, besides many other substances.

This mixture is totally unfit for use and must be purified.

5. *Pass through the hydraulic main.*—From each retort a vertical pipe (*i i i*, Fig. 46) is an outlet for this mixture of gases. This pipe, bending over at the top,

Fig. 46.

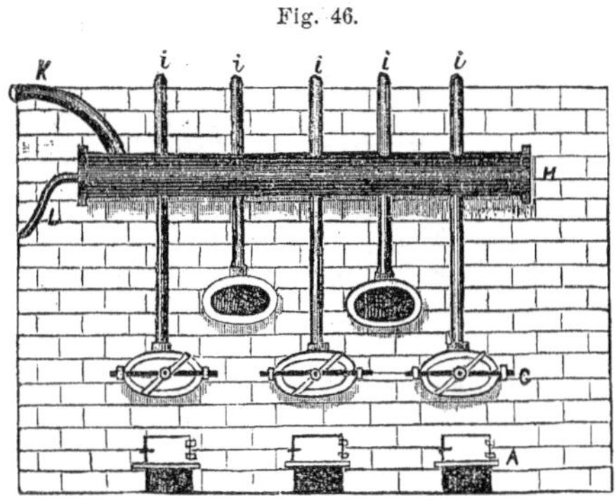

reaches down into a larger and horizontal tube or trunk, H, called the *hydraulic main.* In the beginning of the process this main is filled half full of water, and the pipes *i i*, dip into this fluid. The gas coming over from the retorts bubbles up through the water, which prevents its return. Now the vapors of coal-tar will be condensed in part, by the lower temperature of the main, but as the fluid increases it runs off through the tube L, to a tar cistern. Much of the coal-tar is left in the hydraulic main, while the gas passes out of it through the pipe K.

6. *Through the condensers.*—This pipe, K, leads the gases over to a series of upright pipes C C (Fig. 47), called the condenser. Passing up one and down another until they have traversed the whole series, the gases are

exposed to a large extent of cold surface, and the condensable gases are changed to a liquid state. The tar and ammoniacal liquor thus condensed, runs into a cistern below, from which they may be drawn off at pleasure.

Fig. 47.

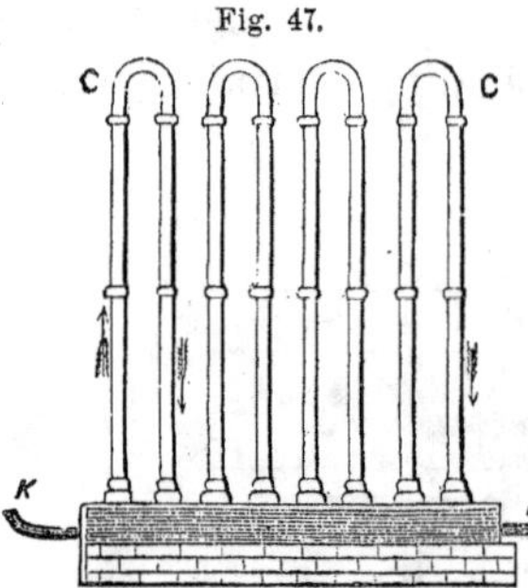

7. *Through the lime purifier.*—The gas still containing sulphur compounds and carbonic acid passes from the condenser through a pipe (L) into a chamber (Fig. 48) in which are several sieve-like shelves, covered with *slaked lime.* In passing through the lime, the gas loses its carbonic dioxide and hydrosulphuric acid.

Fig. 48.

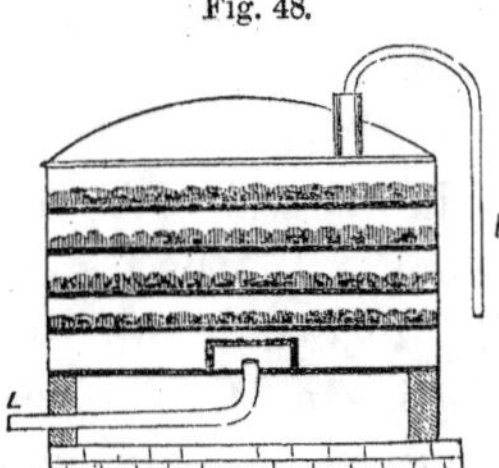

8. *Into the gasometer.*—The purified gas, leaving the lime chamber through a pipe (P), passes into the *gasholder* (Fig. 49), an immense sheet-iron cylinder, closed at the top, and opened at the bottom, hung by chains which run over pulleys at the top, and which carry weights to balance it. Below it is a well of water large enough and deep enough to let this cylinder down until it is filled with water. As the gas enters it, the gasholder rises, and when it is filled, the gas is ready to be pushed out through the pipe (S) into the streets, and finally the houses of the city, furnishing to all a convenient and beautiful light.

"In the iron arteries under towns, in the constellations of burners that rule the nights of favored days, rising over the chaotic oil lamps of old: what a creation!"

(58.) Among the numerous constituents of coal-tar, we may notice carbolic acid and benzole. From benzole, by the action of nitric acid, nitro-benzole is formed; and this, by the action of acetic acid and iron filings, is changed to aniline, from which are made some of the richest colors used in dyeing and calico printing.

Fig. 49.

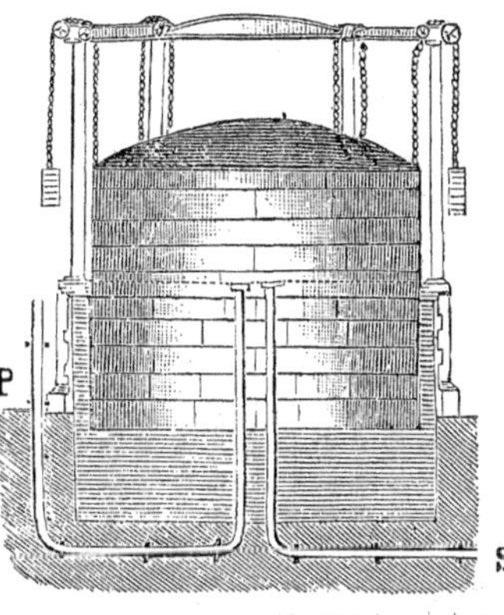

1. *Numerous constituents in coal-tar.*—The coal-tar of the gas-works is a very complex substance. When distilled, vapors containing ammonia first pass over, and then a light oil, known as *coal naphtha*, followed by a heavier one called dead oil, containing a small portion of paraffine. A black *pitch* or *asphalt* is left.

But these are only proximate constituents of the tar: each is itself made up of many simpler ones. The naphtha, for example, is made up of several distinct kinds of oil which may be separated by careful distillation, each having its own particular boiling point.

2. *Carbolic acid.*—One of the most important substances obtained from coal-tar, is carbolic acid ($C_6 H_6 O$). It is not a direct product of distillation, but it is obtained from the naphtha which comes over between 300° and 400° F., by the action of sodic hydrate (caustic

soda). When pure, it is a white solid, soluble in alkalies, with a smell like creosote. Its most important property is its power to interrupt decay: it is a good disinfectant and is in much demand for this purpose. It is used to some extent in medicine, and quite largely in the manufacture of colors for dyeing silk and woolen goods.

3. *Benzole.*—Benzole ($C_6 H_6$) is another important constituent of coal-tar. It is a colorless liquid, sometimes used as a burning-fluid, often, for taking oil-stains from garments: it is a powerful solvent of oils, but on account of its exceeding inflammability it is a dangerous fluid for lamps.

4. *Nitro-benzole.*—By mixing benzole with strong nitric acid a reaction is brought about, shown in the equation,

$$C_6 H_6 + H N O_3 = C_6 H_5 N O_2 + H_2 O.$$

It will be noticed that one atom of hydrogen in the benzole is replaced by one molecule of the radical $N O_2$, forming $C_6 H_5 N O_2$. This new substance is called *nitro-benzole.*

Nitro-benzole is a fluid having the odor of bitter almonds. It is used in making perfumes and confectionery, but its most important use is in the production of aniline.

5. *Aniline.*—Nitro-benzole may be changed to aniline in different ways. Hofmann's method consists in acting upon it by hydrogen set free from sulphuric acid by zinc. The following reaction takes place:—

$$C_6 H_5 N O_2 + 6H = C_6 H_7 N + 2H_2 O$$

losing two combining weights of oxygen, and gaining

two of hydrogen, the nitro-benzole is changed to C_6H_7N. This new substance is aniline.

On a large scale aniline is made by a more economical method (Bechamps'), in which the change is effected by iron and acetic acid. One hundred parts of the crude nitro-benzole is mixed with nearly its own weight of strong acetic acid, and to this is added, little by little, about one hundred and fifty parts of iron turnings. A complicated reaction takes place, and when the mixture is afterward heated, impure aniline is obtained. It is purified by treating it with lime or soda, and redistilling it. By this means the crude aniline of commerce is obtained.

Aniline, when pure, is a colorless liquid, heavier than water, soluble in alcohol and ether, and very slightly in water. Its most remarkable property is that of acquiring various and rich colors by the action of different oxidizing agents. Various rich shades of red, blue, yellow, and other colors, may be obtained from crude aniline. Indeed, almost every variety of tint may be made from the products of coal-*tar!*

II.—DECAY.

(59.) Organic matter buried in the earth undergoes a slow process of destructive distillation. The varieties of mineral coal and petroleum, or rock oil, are doubtless the products of such a process.

1. *Slow destructive distillation.*—The decay of vegetable matter, when buried in the moist earth, or covered by the water and mud of bogs and marshes, is somewhat different from its decay when exposed to the air. Instead of giving off carbonic dioxide and water, and

crumbling to a black mold, it gives off marsh gas and other hydro-carbons, and yields a residue of coal.

The chief constituents of wood are carbon, hydrogen and oxygen—the first two being far the most abundant. On exposure to air, oxygen from the atmosphere combines with hydrogen and carbon of the wood to form water and carbonic dioxide, leaving oxygen in combination with what remains of both these elements, forming humus; but when the air is excluded, the process of decay must consist chiefly in the rearranging of the elements of the wood itself. Its oxygen takes carbon enough to form carbonic dioxide; its hydrogen takes carbon also, and forms gaseous or liquid compounds, while the excess of carbon, not thus used, is left in the form of coal.

2. *Varieties of coal.*—Vast quantities of vegetable matter, accumulating in low wet lands of warm countries, gradually become covered with water, and sometimes, by the sinking of the land, they are buried under mud and sand brought over them by streams or floods. Thus shut off from the air, a slow decay goes on, by which they are at last changed to coal. The different varieties of coal mark the different stages of the process. In peat the change is only well begun; in anthracite coal the process is at an end. The warmth of the earth assists the change, and the great pressure of the material, accumulating for ages upon it, would have much to do with the final compactness of the remaining coal. In bituminous coal the liquid hydrocarbons remain, but may be driven away by heat: from anthracite they have already escaped.

3. *Petroleum.*—Numerous and extensive beds of coal have thus been produced by the slow distillation

of vegetable matter during past ages of the earth's history. But what has become of the liquid hydro-carbons which must have been formed, but which are no longer held in the hard coal? Moreover, during the deposition of other rocks in which no coal is found, there is abundant evidence that plants were growing, and they, too, must have been decomposed in a similar way; what has become of the products of their decay? The gaseous products would, of course, for the most part, escape into the air, and it would be natural to suppose that the liquid products would gradually collect in cavities and fissures in the rocks. Now, inflammable, oily substances, issuing often in large quantities from the fissures of rocks have been long known. To them the general name of *petroleum* has been given. They resemble the liquid products obtained by destructive distillation of wood, and it is believed that they are the products of the slow decomposition of organic matter, chiefly vegetable. Petroleum, or rock oil, is then the liquid hydro-carbon substances given off in the slow process of the decay of vegetable matter long buried in the earth.

The purer varieties of native oil are nearly colorless, and leave no residue when distilled; naphtha is the name given to such. Others are dark colored, and by distillation yield naphtha, and a more or less solid residue called *mineral tar*, or when pure and hard, *asphaltum*. Asphaltum itself is found in many places, the shores of the Dead Sea, Barbadoes, and Trinidad are examples.

From these mineral oils, by distillation, substances very much like those obtained from coal-tar may be obtained, and may be used for similar purposes in the arts.

CHAPTER VI.

ON DECOMPOSITION BY FERMENTS.

General Statement.—Many organic substances are decomposed by the presence of decaying organic matter which contains nitrogen. The action is called fermentation.

I.—THE ALCOHOLIC FERMENTATION

(60.) Sugar, or substances which may be changed to sugar, will be broken up by the presence of ferments into alcohol and carbonic dioxide.

1. *Sugar.*—The sugars are an important class of compounds occurring in the bodies of plants. They are, all of them, compounds of carbon, hydrogen, and oxygen, and in their composition there is this peculiarity—the hydrogen and oxygen are in the proportions to form water. There are many varieties of sugars, but they may be grouped in two classes—the *sucroses* and the *glucoses*

a. *The sucroses.*—Cane sugar, so common and well known, and milk sugar, or lactose, obtained by evaporating the whey of fresh milk, are members of the first class. Cane sugar occurs in the juices of many plants. It is obtained by evaporating the sap of the sugar maple, the juice of the beet, and in far larger quantities from the juice of the sugar-cane. Its general character is well known. Its composition is given in the symbol $C_{12} H_{22} O_{11}$. When strongly heated it yields water and a dark-colored residue, called caramel. One of its most curious properties is shown in its action upon polarized light: it turns the plane of polarization to the right hand.

b. *The glucoses.*—Grape sugar and fruit sugar are glucoses. They are found together in many kinds of fruit, especially in the grape. They have the same composition, $C_6 H_{12} O_6$, and yet differ in several properties. Grape sugar easily crystalizes; fruit sugar never does. The latter is more soluble, and rotates the plane of polarization *to the left*, the former to the right.

When cane sugar is acted on by dilute sulphuric acid, a reaction takes place by which the cane sugar is changed into grape sugar and fruit sugar, by taking the elements of a molecule of water:—

$$C_{12} H_{22} O_{11} + H_2 O = C_6 H_{12} O_6 + C_6 H_{12} O_6$$

Cane sugar + Water = Grape sugar + Fruit sugar.

2. *Substances which can be changed to sugar.*—Among the substances which may be changed to sugar, we may notice starch and dextrine.

a. *Starch.*—Starch consists of a white powder, composed of granules, which have different size and shape in different varieties; those of potato starch being about .007 in. in diameter; of beet root about .0002 in. These granules are not soluble in cold water, but when heated in water, they swell up and split open, and if the paste thus formed is boiled in a larger quantity, the starch is at last dissolved.

When free iodine is brought in contact with starch, a compound, having a rich blue color, is made. This action is the most delicate test of the presence of starch: it will show its presence in potato when the freshly cut surface of the vegetable is washed with a solution of iodine.

The composition of starch is shown by its symbol, $C_6 H_{10} O_5$. By the action of dilute sulphuric acid starch is changed into dextrine and grape sugar.

b. *Dextrine.*—Dextrine, or, as it is more commonly called, British gum, is very soluble in water, and it is used sometimes instead of gum-arabic in calico printing and other arts. It is made from starch, not only by means of dilute sulphuric acid, but by simply heating the starch to about 150° C. (302° F.), or by the action of *diastase* (a substance contained in malt). By the continued action of the diastase the starch is first changed into dextrine and grape sugar, and the dextrine is finally changed also into grape sugar.

$$\underset{\text{Starch}}{3\,(C_6 H_{10} O_5)} + \underset{\text{Water}}{H_2 O} = \underset{\text{Dextrine}}{2\,(C_6 H_{10} O_5)} + \underset{\text{Grape sugar.}}{C_6 H_{12} O_6.}$$

The composition of dextrine and of starch are the same, but their properties are different. Dextrine is soluble in water and is reddened, instead of being turned blue by iodine.

3. *Ferments.*—By the term *ferment* we mean an organic compound, containing nitrogen, and which readily decomposes on exposure to air. Any substance, containing nitrogen and partially decomposed, will act as a ferment. Yeast is the most familiar example. When the sweet juices of vegetables are exposed to the air, a ferment is soon formed in them, and the smallest quantity of ferment being present, an action is started

by it, which goes on until the entire body of liquid is decomposed.

4. *Fermentation.*—The decomposition caused by ferments is called fermentation. It may be easily illustrated by experiment. Dissolve about 100 grs. of honey, or, it may be, molasses, in a pint of water; fill a small flask with the solution, and add a few drops of brewer's yeast. Close the neck of the flask with the hand, and invert it in a dish holding some of the same sirup, and leave it in a warm place for twenty-four hours. Fermentation soon begins; a colorless gas collects in the flask, which, by lime water, may be shown to be carbonic dioxide, while alcohol remains in the fluid.

$$C_6H_{12}O_6 = 2\,C_2H_6O + 2\,CO_2.$$

Sugar = Alcohol + Carbonic dioxide.

All fermentation which produces chiefly alcohol and carbonic dioxide is called the *alcoholic* or *vinous* fermentation. The process goes on best at a temperature of 25° or 30° C. (77° or 86° F.).

(61). Alcohol is the intoxicating principle of all spirituous liquors. From alcohol sulphuric ether is made by the action of sulphuric acid; other acids produce other varieties of ether.

1. *Spirituous liquors.*—The spirituous liquors of commerce, such as brandy, gin, and whisky, are produced by distilling fermented liquids. The fermented liquid obtained from malted grain is called *beer;* that from the juice of the grape is called *wine.* By distilling these and adding various substances to color and flavor the result, different kinds of liquor are made. Brandy

is made by distilling wine; gin is made from different kinds of corn spirits, its flavor being given by juniper berries, sweet flag, liquorice powder, and several other substances. Whisky is also obtained by distilling the fermented liquor from corn.

2. *Alcohol.*—The intoxicating principle in all these liquors is alcohol, which has been produced by fermentation. Distilled from wine, it has been called *spirit of wine.* But mere distillation from the fermented liquor, while it may furnish a concentrated spirit, can not give one entirely free from water. The attraction of alcohol for water is so strong, that a small portion will be retained by it after the most careful distillation. It can be removed by the stronger attraction of quicklime, and when this is done, the product is called *absolute* alcohol; but on exposure to air, it soon absorbs water again, so that absolute alcohol is of rare occurrence. The specific gravity of absolute alcohol is .794: of the strongest commercial alcohol, which contains about 11 per cent. of water, it is .825.

Alcohol is a very combustible fluid, and burns with a pale flame without smoke, producing carbonic dioxide and water. On this account, and because its flame is the source of intense heat, alcohol has been a most valuable fuel to the chemist—his alcohol lamp was formerly in almost constant use; the gas-lamp is now much used instead.

(62.) By the action of concentrated sulphuric acid upon alcohol, sulphuric ether is formed; other acids yield different varieties of ether.

1. *Sulphuric acid upon alcohol.*—When equal weights

of concentrated sulphuric acid and alcohol are heated to about 140° C., a very volatile substance is formed, whose vapors may be condensed in a separate vessel. It is *sulphuric ether*, or, since it is the only ether of commercial importance, it is called simply *ether*. The chemical change is complicated, but the result of it all is that the alcohol is changed into ether and water.

$$2\,(C_2\,H_6\,O) = C_4\,H_{10}\,O\,(\text{ether}) + H_2\,O.$$

2. *Ether*.—Ether is a transparent liquid, with a peculiar odor, and a sweetish taste. When breathed it causes exhilaration at first, but perfect insensibility at last. On this account ether has been used to render patients insensible to the pain of surgical operations.

The evaporation of ether produces intense cold: a pretty experiment can illustrate this. Let a drop or two of water be covered with a few drops of ether, and by a bellows or a blow-pipe, a current of air be blown against the ether. By its rapid evaporation the ether takes away the heat, until the water is frozen. A mixture of ether with solid carbonic dioxide will produce a temperature of −110 C°. (−166° F.). Pure ether has never been frozen.

Ether is very combustible, burning with a bright flame; and a mixture of its vapor with air is explosive.

Ether is used in medicine, and extensively by the chemist as a solvent. Oils and resins, caoutchouc, and many other organic substances are soluble in this liquid. It is a more powerful solvent than alcohol.

3. *Varieties of ether*.—In chemistry the term ether is given to a class of volatile compounds produced from alcohol by the action of strong acids. By the

use of nitric acid, *nitric ether* is made; and other ethers by other acids. Thus hydrochloric acid will produce *hydrochloric ether*.

4. *Ethyl.*—Ether may be considered as a compound of oxygen with two molecules of the radical C_2H_5—called *ethyl*. Its symbol,

$$C_4H_{10}O, \text{ is equivalent to } \left.\begin{matrix} C_2H_5 \\ C_2H_5 \end{matrix}\right\} O,$$

and according to this the true name for ether is *ethylic oxide*. This radical, C_2H_5,—ethyl, is contained in a numerous class of compounds, known as the ethyl series. Alcohol itself is a member of this series; its symbol

$$C_2H_6O \text{ being equivalent to } \left.\begin{matrix} C_2H_5 \\ H \end{matrix}\right\} O,$$

and its name would accordingly be *ethylic hydrate*. So, too, with chlorine this radical would form ethylic chloride (C_2H_5Cl); with potassium, potassic ethylate (C_2H_5KO).

Ethyl has been described by Dr. Frankland as a very heavy, colorless gas, with a slight odor of ether, burning with a bright flame. It is of no practical importance in the arts, but to the chemist its theoretical value is very great, simplifying, as it does, his study of the composition of a long series of organic substances. And this is only one of many such radicals, out of each one of which the symbols of an entire series of organic substances may be said to grow.

Some of these radicals, with their corresponding alcohols, follow in the table:—

NAME.	SYMBOL.	ALCOHOL.	SYMBOL.
Methyl	$C H_3$	Methylic Alcohol	$C H_4 O$
Ethyl	$C_2 H_5$	Ethylic "	$C_2 H_6 O$
Propyl	$C_3 H_7$	Propylic "	$C_3 H_8 O$
Butyl	$C_4 H_9$	Butylic "	$C_4 H_{10} O$
Amyl	$C_5 H_{11}$	Amylic "	$C_5 H_{12} O$

Each alcohol is but a first step in a long series of organic compounds.

II.—THE ACETOUS FERMENTATION.

(63.) An alcoholic liquid, containing a small quantity of a ferment, and in presence of air, yields vinegar. Vinegar consists chiefly of water and acetic acid.

1. *Production of vinegar.*—When an alcoholic liquid is exposed to the air in a warm place, a little yeast or other nitrogenous matter in it will start an action by which the alcohol is changed into vinegar. "A good extemporaneous vinegar may be prepared by dissolving one part of sugar in six of water, with one part of brandy and a little yeast. The mixture is put into a cask, with the bung-hole open, and kept at a temperature between 70° and 80° F. In from four to six weeks the clear vinegar may be drawn off." (Brande & Taylor.) A still more simple process consists in soaking apple skins in soft water for a few days; straining the juice, and letting it stand exposed to the air in a warm place for several days: an excellent vinegar is the result.

2. *Acetic acid.*—Common vinegar is composed chiefly of water and acetic acid. Its quality depends upon the proportion of acid it contains, and the absence of other

impurities. The composition of acetic acid is shown by the symbol $C_2H_4O_2$. It is a colorless liquid, with a powerful and peculiar odor, which, once experienced, is afterward easily recognized.

3. *Fermentation of alcohol.*—The chemical action by which alcohol is changed to acetic acid is called the *acetous fermentation.* And yet it is not in all respects a true fermentation. It is not a *decomposition,* but rather an *oxidation,* as may be seen by comparing the symbols of alcohol and acetic acid. In this respect the action is a case of combustion rather than of fermentation. It can take place only in the presence of air, so that the action is not entirely due to a ferment. Yet, on the other hand, it will not occur, except in presence of a nitrogenous substance, to which the term *ferment* has been given; and hence the reaction is very naturally called a fermentation.

CHAPTER VII.

ON THE CHEMICAL ACTION OF LIGHT.

(64.) The combination of hydrogen and chlorine, when a mixture of the two gases is exposed to light, has been noticed: other examples of the chemical action of light are to be described. The rays which produce these effects are, in general terms, the most refrangible part of the beam.

1. *Hydrochloric acid formed under the influence of light.*—We have seen (p. 49) that when a mixture of hydrogen and chlorine is exposed to diffuse light, a gradual combination takes place, and hydrochloric acid is formed. We may now study this action more fully. Let a strong tube or bottle be filled with a mixture of the two gases, taking care to have a little more of one—say of hydrogen—than of the other. Let this mixture be prepared in a dark room, the tube containing it being inverted in a vessel of water, firmly fixed in place (Fig. 50), and covered with a black cloth. Thus prepared, take the apparatus at once to a place where the *direct rays* of the sun may fall upon it, and by means

Fig. 50.

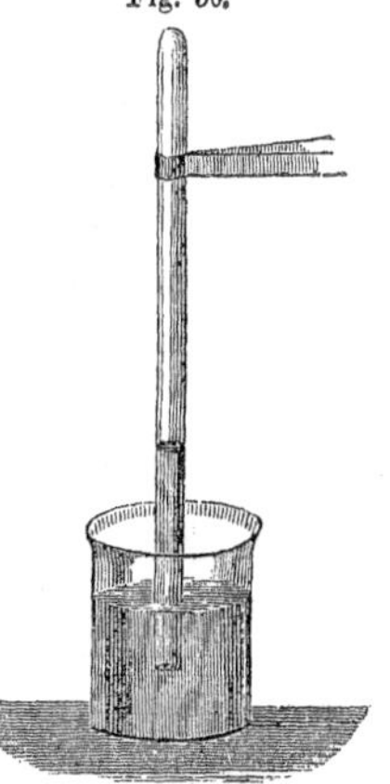

of a long handle remove the cloth. A violent explosion will quickly follow: the water, speedily dissolving the acid gas, will rise in the tube, and would strike the top of it with violence, did not the excess of hydrogen act as a cushion to prevent it.

Exposed to *diffuse* light, the combination of the mixed gases is gradual, but if prepared and kept in the dark, no combination occurs.

These experiments clearly illustrate the fact, that sunlight has power to produce chemical action. Other intense lights, electric light, for example, may be used with similar results.

2. *Effect of light upon silver compounds.*—Into a test tube put a quantity of a solution of argentic nitrate (nitrate of silver), and add a few drops of hydrochloric acid. A heavy white precipitate of argentic chloride is at once formed, which speedily settles to the bottom. Let the tube be now placed in the sunlight, and an interesting change of color will be gradually produced. The snow-white chloride becomes pink, violet, brown, and at last dark bronze-black. The chloride, dry and pure, will not show this change on exposure to light: moisture seems to be necessary. Light causes a reaction between water and the chloride, by which hydrochloric acid is formed, and a small quantity of metallic silver is set free. To the presence of this metal in a state of very fine division the darkening is due.

Argentic nitrate is not changed by the action of light unless in contact with organic matter: it then blackens like the chloride.

When argentic iodide is exposed to light, no *visible* change occurs, but still a molecular change does take place. This curious fact may be shown by experiments.

Let two highly polished plates of silver be held in iodine vapor in the dark; by this means a thin film of iodide will be made over their surfaces. Leave one plate in the dark, and put the other, for a time, in the sunlight: the two plates still look alike: the light has caused no visible change. But now hold both, in a box at moderate temperature, over the vapor of mercury: the one which was exposed to the light immediately blackens; the other is not changed. The darkening in this case is due to a combination of mercury with the silver of the iodide—the mercury decomposing the iodide after exposure to the light. The nature of the action of light in this case has been in doubt. The best explanation (Amer. Jour. of Sci., vol. 42, p. 198, and vol. 44, p. 71) supposes it to be a *physical* action, by which the molecules are so disturbed as to yield afterward to the attraction of mercury.

3. *Effects of light upon other compounds.*—Some of the compounds of several other metals are easily affected by light. Solutions of gold in contact with organic matter yield metallic gold; and the compounds of iron, mercury, uranium, and some others, are either reduced to a lower state of oxidation, or, in rarer cases, to the metallic state.

Many chemicals kept in the light in the laboratory, are in time sensibly changed. Phosphorus is changed to the *red* allotropic form; and nitric acid becomes slowly yellow by the presence of lower oxides of nitrogen into which it is partially decomposed.

4. *Effect of light upon the vapors of volatile liquids.*—Some new and very interesting effects of light have been lately pointed out by Dr. Tyndall. (See Chem. News, Amer. Rep., vol. 4, p. 65.) The vapors of vari-

ous liquids have been acted upon by concentrated light in a glass tube, with curious and beautiful results. Among them are the following, obtained by experiment with the organic compound—amylic nitrate. The tube being filled with mixed air and vapor, a beam of electric light was sent through it from end to end. "For a moment nothing whatever was seen within it; but before a second had elapsed, a shower of liquid spherules was precipitated on the beam, thus generating a cloud within the tube. This cloud became denser as the light continued to act, showing at some places vivid iridescence."

"The beam of the electric lamp was now converged so as to form within the tube, between its end and the focus, a cone of rays about eight inches long. The tube was cleansed, and again filled in darkness. When the light was sent through it, the precipitation upon the beam was so rapid and intense, that the cone, which a moment before was invisible, flashed suddenly forth like a solid luminous spear."

"When the vapor was permitted to enter the tube unmixed with air or any other gas, the effect was substantially the same. Hence the seat of the observed action is the vapor itself."

"I have taken no means to determine strictly the character of the action here described, my object being to point out to chemists a method of experiment which reveals a new and beautiful series of reactions; to them I leave the examination of the products of decomposition."

5. *The rays that produce chemical action.*—Not all the rays of the sunbeam take part in these chemical actions. If a sheet of white paper is washed with a solution of argentic nitrate, and then used as a screen to re-

ceive the solar spectrum, it will be darkened, but not alike in all the colors. The least effect will be found in the yellow—indeed a long time will be needed to darken the paper there at all. In the red rays a slight change will be noticed, while on the colors toward the other end of the spectrum the change is very decided. And what is more curious, the effect is seen upon the paper, *beyond the violet*, where there is no light. By letting these dark rays pass through the screen that stops the colored light, we may let them fall upon the prepared paper *in the dark*, and we shall find that it is blackened by them. Luminous rays are not necessary to chemical action.

We therefore conclude that the sunbeam contains a set of *invisible* rays, by which its chemical action is produced. These chemical rays are distributed through all parts of the solar spectrum, but most abundantly in the violet end, and points just beyond it, and hence we may in general terms say that they are more refrangible than the luminous rays.

(65.) The beautiful art of photography depends on the chemical action of light chiefly upon the compounds of silver. We may notice, briefly, the process: 1st, upon silver; 2d, upon glass; and 3d, upon paper.

1. *Photography on silver.*—Photography is the art of making pictures by means of light; and all pictures produced by the action of light are photographs. The daguerreotype is a photograph on silver.

The following is an outline of the process of making the daguerreotype. A highly polished surface of silver, usually a plate of copper, silver-coated, is exposed to the vapors of iodine and bromine, in a dark box. By this means a thin film of argentic iodide ($Ag\ I$) and bromide ($Ag\ Br$) is formed by direct combination. This film has a bronze-yellow color, and is very sensitive to light. The plate is then put into a camera, where it receives the light from the object to be pictured, and after a few seconds' exposure, it is taken out into a dark room. The same bronze-yellow surface, with no sign of a picture, remains,—no *visible* change having been made by the light. The plate is next held in the vapor of mercury, when an image immediately appears, due to an amalgam of mercury and silver formed on those parts which have received the light, but not upon others. This amalgam can not be washed off, but the undecomposed bromide and iodide is dissolved and washed away by a solution of sodic hyposulphite, which is next poured over it. The highly polished silver beneath forms the deep shades, while the light gray amalgam forms the lights of the picture. A film of gold is then spread over the picture by pouring upon it a dilute solution of auric chloride (chloride of gold), and heating it; and finally, a thorough washing completes the operation.

The daguerreotype, better than other processes of photography, can bring out the minute details of objects. "In 1846 we obtained by this process a copy of the 10,000 letters of the Greek inscription on the Rosetta stone of the British Museum, within the space of two square inches." (Brande and Taylor.) The disadvantages of this form of picture are, that the image can not be easily seen in all lights, and it is liable to tarnish on exposure to air.

2. *Photography on glass.*—Glass may be coated with a film of substance sensitive to light, and a photograph may then be made upon it.

Cotton, acted on by a mixture of nitric and sulphuric acids, is changed to gun cotton, or pyroxyline, and this, when dissolved in alcohol and ether, forms the substance called *collodion*. A small quantity of an iodide (potassic), or a mixture of two or three, is put with the collodion, and the solution is then rapidly poured from a wide-mouthed vessel over the very clean and dry surface of the glass; the alcohol and ether quickly evaporate, and leave a thin film upon the glass, which is then plunged into a bath of argentic nitrate. While in this bath for a few minutes, the iodine in the film takes silver from the nitrate, and forms argentic iodide (Ag I). The plate being then taken from the bath, thoroughly washed and dried, is ready to receive the action of light.

The plate, with the sensitive film on its surface, is exposed in a camera for a short time, during which the light from the object performs its curious action upon the iodide, by which no picture is made visible, but by which one is prepared to be developed. On pouring over the film of iodide a solution of pyrogallic acid ($C_6 H_6 O_3$) or of ferrous sulphate ($Fe S O_4$), containing a few drops of argentic nitrate, the picture is brought out or "developed," as it is technically described.

The next step is to dissolve away the undecomposed iodide. This is done by washing the plate in a saturated solution of sodic hyposulphite. After this, the plate is thoroughly washed, dried, and varnished on the picture side, to preserve the film from injury. Often a black varnish is put upon one side of the plate to serve as a background to the picture; it forms the shades of the picture, while the portions where the action of light was strongest, and hence the most silver deposited, furnish its lights. In other cases a piece of black cloth put behind the picture furnishes its shades.

3. *Photography on paper.*—When a photograph on glass is viewed by transmitted light, it will be seen that the shades of the object are lights on the picture—the lights and shades being inverted. Such a picture is called a *negative.* If the "development" be pushed further, more of the iodide in the film will be decomposed, and the silver will more or less completely intercept the light. From such a negative a picture may be printed upon paper.

The paper for this purpose is prepared by floating it first upon a solution of salt, and afterward upon one of argentic nitrate. Argentic chloride ($Ag\,Cl$) is thus formed in the paper. This paper, placed under the negative, is exposed to light; the light is stopped by the shades of the negative, but passes freely through its lights; upon the paper, therefore, the lights and shades correspond with those of the object from which the negative was taken. Such pictures are called *positives.*

The photograph is next to be thoroughly washed and then "toned." The toning-bath is a solution of hydrosodic carbonate (bicarbonate of soda), containing a little auric cloride ($Au\,Cl_3$). The change produced by this bath is beautiful to witness. From an unpleasant, dull, reddish color the picture visibly changes to a rich blue-black. The picture, having next been washed in water to remove all traces of the gold bath, is soaked for some minutes in sodic hyposulphite, to wash out all of the undecomposed chloride, and afterward for twenty-four hours in water, to remove all traces of the hyposulphite.

Every part of the photographic process needs the

greatest care and skill. "The causes of failure at every stage are numerous, and are sometimes difficult to explain."

(66.) The solar spectrum, viewed through a spectroscope, is seen to be crossed by a host of fine, dark lines, but the spectra of burning elements, viewed in the same way, are crossed by bright lines. Now, the bright lines may be changed to dark ones by absorption, and the solar lines are thought to be dark by absorption in a similar way.

1. *The spectroscope.*—A spectroscope is an instrument by which to observe the lines which cross the spectra of natural or artificial light.

Fig. 51.

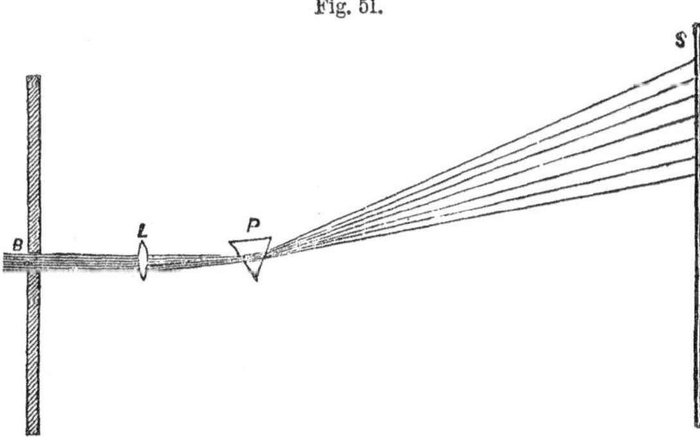

A beautiful experiment illustrates the principle of the spectroscope. A beam of sunlight, B (Fig. 51), after entering a darkened room through a narrow slit, is passed through a convex lens, L, and then through a prism, P, of carbonic disulphide ($C\ S_2$). The spectrum, falling upon a white screen, S, will be seen crossed by several dark lines (Fig. 52). This effect may be seen by many persons at once; but if a telescope be pointed toward the prism, one person, looking though it, may see the lines in far greater number and much more

Fig. 52.

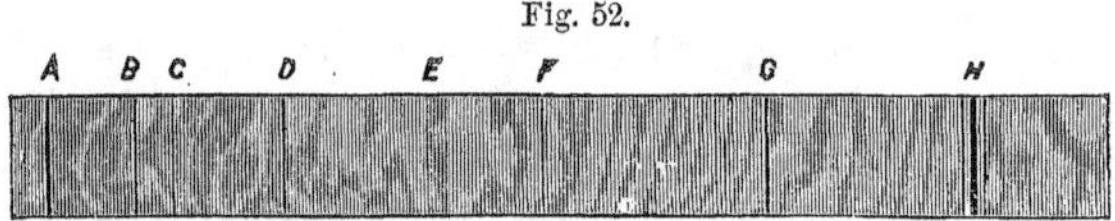

distinct. Now the slit and lens may be fixed in a tube, and this, with the prism and telescope fixed in proper positions on a stand, constitutes a spectroscope. In the instrument shown in Figs. 53 and 54, the light enters the

Fig. 53.

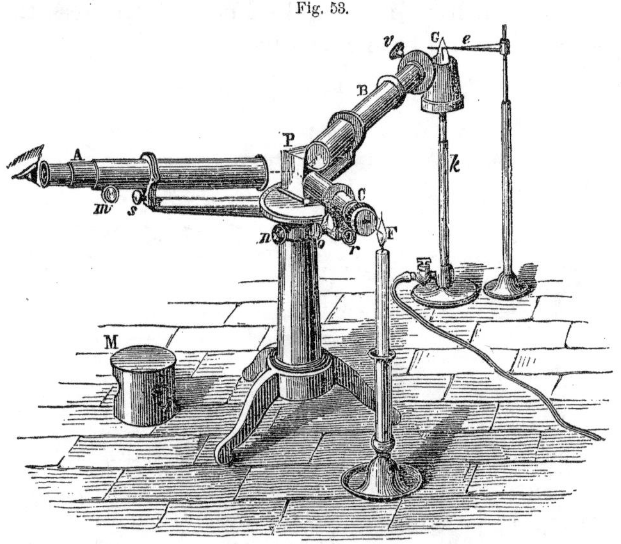

slit at one end of the tube, B, and passing through a lens at the other end, falls upon the prism, P. The spectrum is viewed by looking through the telescope, A.

2. *The dark lines.*—The dark lines of the solar spectrum to be seen by means of a spectroscope are very numerous; the existence of several thousand has been ascertained. So fixed are the positions of these lines, that the spectrum has been mapped—the invariable place of each line being shown according to a scale of equal parts. For this purpose, a small scale, very accurately graduated, is placed in a third tube, C (Figs. 53 and 54), and illumined by a light, F. The light from this scale is reflected from the face of the prism, P, into the telescope, A, so that the image of the scale will be seen alongside of the spectrum, and the place of each line may be marked by the divisions of the scale.

Several of these dark lines, first mapped, were named by Fraunhofer from letters of the alphabet.

Fig. 54.

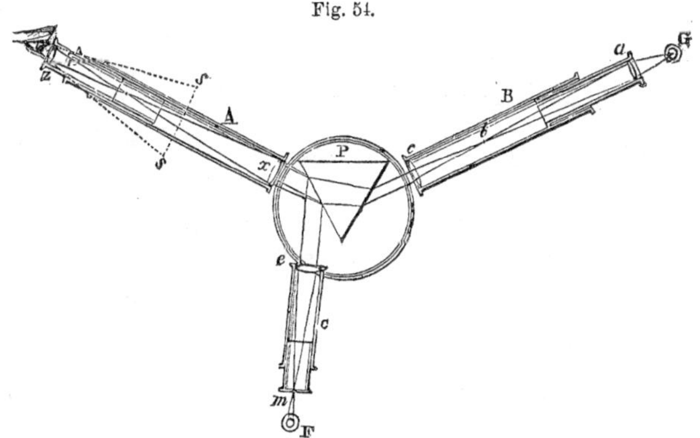

The line A lies near the end of the red color; B in the middle, and C at the boundary between the red and orange. D is in the yellow; E in the green; F in the blue; G in the indigo, and H in the violet.

The light of the sun always gives the same set of lines, but starlight gives a different set, and each star seems to have a set of its own.

3. *The bright lines.*—When artificial light is used, the spectroscope reveals sets of very brilliant lines, whose position and color differ when different substances are burned. Each element seems to furnish a set of its own so characteristic, that, as we have seen (p. 31), the spectrum may be relied upon to show the presence of different substances, a yellow line, Fig. 56 (two, as seen by a powerful instrument), indicating sodium in the flame, while crimson bands in the red end and a blue line in the blue end of the spectrum, declare the presence of strontium. In Fig. 55 several of these spectra are

Fig. 55.

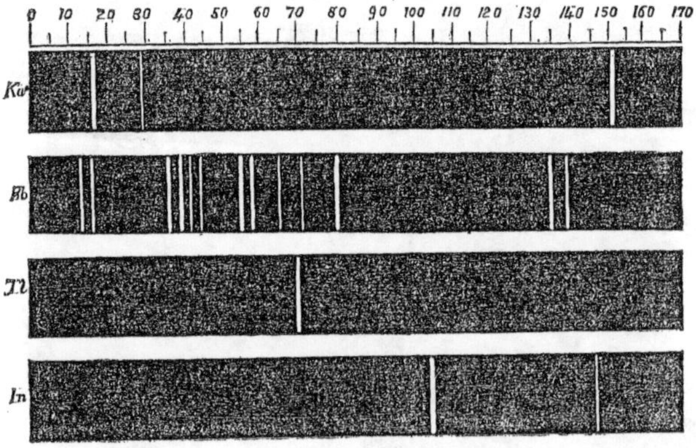

represented, showing the *position* of the characteristic lines—first of potassium (Ka); second of rubidium (Rb); third of thallium (Tl); and fourth of indium (In): no attempt is made to represent the colors. Of potassium, one line is red, the other violet; of rubidium, the two at the left are crimson, the two at the right are blue, while those between are yellow and green; of thallium the single line is green; of indium the characteristic line is blue. By their spectra the last three metals named were first discovered; cæsium was discovered in the same way.

The alkaline metals are very easily detected by their spectra. They are easily burned in a Bunsen's burner, and each gives a small number of lines. Others may be examined with greater difficulty, because less volatile, and because of the greater number of lines they furnish. Iron, for example, needs the electric heat to vaporize and burn it, and there are at least eighty lines in its spectrum.

4. *Bright lines changed to dark ones.*—If, while a sodium flame is giving its peculiar yellow lines, the lime light is placed behind it, so that its rays, going through the spectroscope, fall upon the sodium spectrum, the brilliant yellow lines are at once changed to dark ones. This wonderful effect is produced by absorption and contrast: *absorption*, since the yellow sodium vapor absorbs those rays of the lime light which would fall upon the spectrum where the yellow lines are formed; and *contrast*, since on both sides of the yellow lines there is the many times stronger light of the lime, in contrast with which the yellow lines *appear* dark.

The brilliant lines from potassium, barium, and other metals, have been changed to dark ones in the same way—that is, by sending through their burning vapors the rays of some intenser light. The explanation is generally given by saying that "a luminous vapor will absorb rays of the same refrangibility as those which itself emits" and adding that "the lines are bright or dark, according as they are lighter or darker than the adjacent parts of the spectrum."

5. *Solar lines dark by absorption.*—We find ourselves able, then, to imitate, in a degree, the lines of the solar spectrum: can it be that our *method* is an illustration of the way in which the solar lines are produced? Such is the theory. It is thought that an immense and luminous atmosphere surrounding the sun contains the vapors of various substances whose light alone would give bright lines, but that the far more intense light of the orb itself coming through this atmosphere renders them dark by contrast.

This theory is curiously confirmed by observations lately made. The "red flames" or "prominences" seen during eclipses of the sun have been examined, and, while the body of the sun was in total eclipse, they gave a spectrum with *bright lines.* This observation was made by M. Janssen and others upon the total eclipse of the sun in 1868. About the same time Mr. Lockyer, in England, by means of a superior instrument, obtained a spectrum with bright lines, in clear daylight. By pointing his instrument toward the *edge* of the sun, and then sweeping around it, on coming to a prominence, its spectrum was crossed by three brilliant lines. The atmosphere of the sun then, alone, gives bright lines: when shone through by the stronger light of the orb, the lines are dark. (See Chem. News, Am. Rep., vol. 4, p. 19.)

(67.) Certain bright lines of artificial spectra coincide exactly with the black lines in the solar spectrum, and hence they must be caused by the same substance. On this principle we may learn something of the composition of the heavenly bodies.

1. *Coincidence of dark and bright lines.*—"It is easy to construct the spectroscope so that the two halves of the slit may be illuminated from different sources. If then, we admit a beam of sunlight through one half and the light of a sodium flame through the other half, we shall have the two spectra side by side in the same field, as in Fig. 56, and it will be seen that the sodium

Fig. 56.

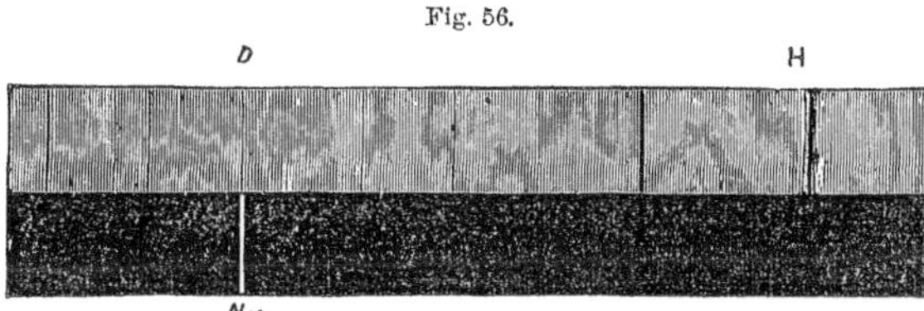

line, which appears as a double line under a high power coincides absolutely in position with the double dark line D, in the solar spectrum." (Cooke.) The coincidence is very striking where the bright lines are more numerous: the eighty bright lines of iron correspond in position *exactly* with eighty dark lines of the solar spectrum.

2. *Caused by the same substance.*—Now, since no two substances have been found to give lines in exactly the same place in the spectrum, and since the same element will give its lines *always* in the same position, it follows that if a set of dark lines and another set of bright lines *exactly coincide*, they must be caused by the same element.

3. *The composition of the sun.*—And if this is true, then sodium and iron, whose lines do thus coincide with dark lines from the sun, must be elements in that body. Other sets of dark lines in the solar spectrum

coincide exactly with sets of bright lines in spectra of elements: we may suppose that all such are constituents of the sun. Several metals have in this way already been detected in that far-distant orb! Besides iron and sodium, there are zinc, copper, nickel, chromium, magnesium, barium, and calcium. Hydrogen has also been detected. *Something* of the composition of the sun is therefore determined.

4. *Composition of the stars.*—On examining the spectra of different stars, we find that each has some lines not found in others, and this suggests that they are not all alike in composition. By comparing the spectrum from the star Aldebaran with the spectra from substances here, several sets of lines are found to coincide. Among our elements thus found to exist in Aldebaran are iron, sodium, mercury, and arsenic. The lines of mercury and arsenic are not yet found in the solar spectrum.

Several other stars have been examined, and something of their composition determined.

5. *Composition of nebulæ.*—In various parts of the heavens are to be seen with a telescope curious bodies which look like patches of light or luminous clouds. When seen through the most powerful instruments, some of these *nebulæ*, as they are called, look like clusters of stars, others have never been seen except as luminous clouds. On viewing these nebulæ with a spectroscope, several of them show the phenomenon of *bright lines.* It would seem that such must be composed entirely of *glowing gas*, because a luminous solid or liquid body enveloped in an atmosphere would give dark lines by absorption. Much doubt, however, still rests upon this interesting subject: indeed this

branch of chemistry is still in its infancy. (See Chem. News, Am. Rep., vol. 1.)

When the dark lines of the solar spectrum were first discovered by Wollaston in 1802, the phenomenon was a thing so very delicate and so apparently useless, that for a long time it received little attention; and had not Fraunhofer rediscovered it several years afterward, and mapped the most noticeable among the lines, the whole thing might very possibly have been forgotten. Little more than fifty years have passed, and yet to what wonderful results has the study of these lines led! The sun and the distant stars are being analyzed; and even the nebulæ, whose distance defied the most powerful telescope, are yielding their secrets to the spectroscope.

CHAPTER VIII.

ON THE CONSERVATION OF FORCE.

(68.) Chemical attraction, heat, light, magnetism, and electricity, are different manifestations of force,* mutually and intimately related.

1. *Chemical attraction.*—That which holds atoms together in chemical combination is as truly a force as that which holds a rock in its place upon the earth, or that which pulls a train of cars. We know it to be a force, because it will do that which it requires force to accomplish. For example: Chlorine gas, at 60 °F., by a pressure of four atmospheres—60 *lbs. to a square inch*, is condensed to a liquid. By no amount of pressure has it ever been condensed to the solid state. But when combined with sodium, it forms a solid without pressure. Hence the chemical attraction here is more powerful than the greatest pressures that have ever been applied to condense the gas.

For another illustration, we may notice that oxygen has resisted all mechanical forces applied, in the hope to condense it even to a liquid: the same thing is true of hydrogen. But under the influence of chemical attraction, they unite to form a liquid, water, without pressure. And more: untold volumes of oxygen are reduced

* Force,—that which is expended in producing motion, either of masses or of molecules.

to a *solid* state, and locked up in the rocks. Surely the influence which causes combination and exerts a power so much greater than any mechanical force that we can exert must be itself a force.

2. *Heat is a force.*—That which by producing atomic motions, causes the sensation of warmth, is also a force. A bar of iron, whose section is $\frac{1}{10}$ of a square inch, may be stretched $\frac{1}{10000}$ of its length by hanging to the end of it a weight of *one ton.* But to raise its temperature 16° F. will lengthen it the same amount. The force which expands it is equal to the force exerted by the one ton weight, and yet it is nothing more than the repulsive force of heat.

3. *Light is a force.*—Admitting that the constituents of compounds are held together by a force, it will not be difficult to see that even light is a force also. Paper, moistened with argentic chloride, and exposed to light, is quickly blackened, showing that the chloride is decomposed. In this case light overcomes the chemical attraction, and since force can be overcome by force only, it follows that light is a force.

4. *Electricity and magnetism.*—Electricity and magnetism are forces. The first shows its power by drawing a pith ball toward an electrified body, or by producing other motions. The second shows itself a force by lifting a piece of iron brought near to the poles of a magnet.

5. *Heat, light, and chemical force intimately related.*—Fixing our attention upon the three forces, heat, light, and chemical attraction, their intimate relation will easily appear; in the first place, from the fact that they so frequently accompany each other. Combustion is the effect of chemical attraction, but it never oc-

curs without heat, and when rapid, never without light. The trio bear each other company, not only in combustion, but also in the sunbeam. The beauty of the solar spectrum, described in natural philosophy, can not be forgotten, but over the visible spectrum only one of three kinds of energy in the sunbeam is spread. *Invisible* heat-rays and chemical rays are also spread upon the screen. Notice Fig. 57, in which these three

Fig. 57.

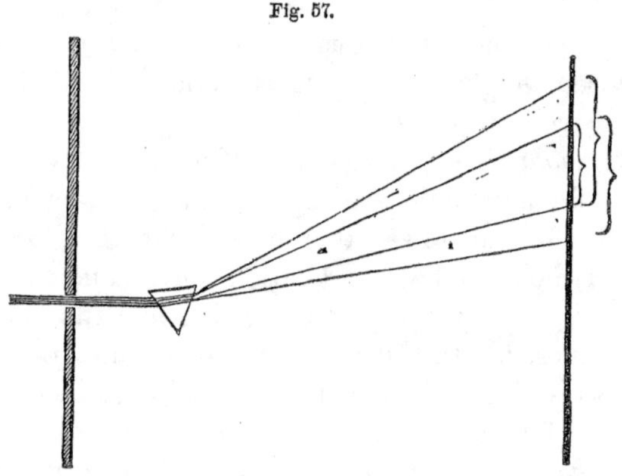

sets of rays are represented, the parts of the screen covered by each being shown by the brackets. In the visible part of the spectrum all the parts may be detected, while the heat rays reach beyond the red, and the chemical rays beyond the violet.

But a closer relation than that of mere companionship is suggested by this threefold spectrum. In general terms, we may say that the chemical rays are more refrangible than those of light, the heat rays less; so that, in this respect, they differ just as the colors violet, yellow, and red, differ from each other. But in natural philosophy, we have been taught that the colors of light depend upon the rapidity of the vibrations which produce them, the *most refrangible* being caused by the *most rapid* vibrations. Violet is more refrangible than yellow, because its vibrations are more rapid; and yellow more refrangible than red, for the same reason. But chemical rays are more refrangible than light; is it because they are in more rapid vibration? So heat-rays being less refrangible than light, are they in less rapid vibration?

That the natures of the different colors are the same has never been doubted, and since heat rays and chemical rays differ from light rays only as one color differs from another, we may suppose that the natures of heat, light, and chemical attraction are the same. The wave theory of light has been accepted; but if, as this theory assumes, the phenomena of light are due to vibrations those of heat and chemical attraction must be due to vibrations also.

6. *Electricity and magnetism related to other forces.*—That electricity and magnetism are closely related to the forces just considered is shown by the fact that the two sets mutually produce each other. In the battery, for example, chemical force produces electricity, and this electricity may produce a vivid light, heat a wire, or magnetize a bar of iron. Heat also may produce electricity, and the electricity may in turn cause a chemical action.

7. *All these forces different forms of a single influence.*—In natural philosophy (Cooley's, p. 31), we have been taught to regard gravitation, cohesion, adhesion, and capillary force as so many different manifestations of a single influence. In like manner, considering the inti-

mate relation between heat, light, chemical force, electricity, and magnetism, the scientist is inclined to believe them all to be only so many different ways, in which a single influence shows itself. His experiments have, moreover, at length brought him to the still more definite truth known as the *conservation of force*. The following is the statement of this most important principle.

(69.) Force, like matter, is indestructible. Its manifestations may change from one form to another, but the same amount which in any form disappears, must reappear in others.

1. *Force is indestructible.*—That matter is indestructible is now well known. In very early times it was not. He who first asserted this truth was compelled to show what became of bodies when they disappear, as when wood "burned up" or water "boiled away." Now forces act and disappear; what becomes of them? Certainly this question must be answered before we may admit that force is indestructible. When matter disappears, it simply changes form; so when forces disappear, while acting, they simply change from one form to another. If this can be proved, then force, like matter, is indestructible.

2. *Illustrations of changes in the manifestations of force.*—Abundant facts in natural philosophy and chemistry, when rightly understood, show how force may flit from one form to another

An iron nail is warmed by the blow of a hammer. In this case, the *force* applied to the hammer produces *motion*, but when the nail is struck, this motion of the hammer stops, and, it *seems* as if the force also ceased to act. But the *heat* in the nail shows that the force of the falling hammer has produced vibrations of the molecules of the nail, and by these vibrations the existence of the force is shown. The force is no longer in the hammer, it is in the nail; it is no longer muscular force, nor gravitation; it is heat.

For another example, let us link together the following familiar facts.

The force of exploding gunpowder speeds a bullet: in this case, *force produces motion.* The bullet strikes a rock and is partly melted by the blow; in this case, *motion checked appears as heat.* Heat caused by the blow of the gun-hammer exploded the percussion cap, in the beginning: in this case *heat expended reappeared as chemical force.*

Or we may illustrate by another series of experiments. *Heat* may reappear as *electricity;* it does, when applied to the junction of two bars of different metals, or to the end of a thermo-electric pile. *Electricity* may reappear as *light;* it does, when a strong current acts through charcoal points. *Light* may reappear as *chemical force*, as when it acts upon a mixture of hydrogen and chlorine. *Chemical force* may reappear as *heat*, as it does in all cases of combustion. Starting with heat, we may thus chase the influence from form to form, until we bring it back to heat again.

Finally, to complete this series of illustrations, a helix, a galvanometer, and a fine iron wire, may be put in different parts of the same battery circuit. *Chemical force* in the cells of the battery reappears as *electricity* in the wires of the circuit, as *magnetism* in the helix, as *heat* and *light* in the fine wire, and as *motion* in the galvanometer needle. The almost instantaneous ap-

10

pearance of this circle of forces is very striking and convincing.

3. *The same amount must reappear.*—But when we are convinced that force may change from one form to another, we are still not sure that it is indestructible. We must go further, and prove, if may be, that the change occurs in definite quantities.

By most convincing experiments, Mr. Joule has shown that a *one pound weight* falling a distance of 772 *feet*, and suddenly stopping, evolves heat enough to raise the temperature of 1 *lb. of water* 1° *F.* The mechanical force in the blow is definite; the amount of heat produced is equally definite. The *forces* of the blow and of the heat it evolves may be considered *equivalent*, so that the force of 1 lb. falling 772 feet, or which is the same thing, of 772 lbs. falling 1 foot, is called the *mechanical equivalent of heat.* Under all circumstances the amount of heat generated by the same amount of force is fixed and invariable.

"A ton of coal, by its combustion, yields a certain definite amount of heat. Let this quantity of coal be applied to work a steam-engine, and let all the heat communicated to the machine and the condenser, and all the heat lost by radiation and by contact with air, be collected. It would fall short of the amount produced by the simple combustion of the ton of coal, and it would fall short of it by an amount exactly equivalent to the quantity of work performed. Suppose that work to consist in lifting a weight of 7,720 lbs. a foot high; the heat produced by the coal would fall short of its maximum by a quantity just sufficient to warm a pound of water 10° F." (Tyndall.) The heat needed to lift 7,720 lbs. one foot, is just the same that would be produced by 7,720 lbs. falling one foot, which is 10°.

The same inflexible law of equivalence holds good in all the changes of force from one form to another. Electricity decomposes chemical compounds, but a certain amount of electricity can cause a fixed or definite amount of decomposition. The relation between electricity and chemical force was discovered by Dr. Faraday. He found that by consuming 65 grs. of zinc by chemical action in the battery, a quantity of electricity was produced, which, passed through water, set free 16 grs. of oxygen. Now 65 and 16 are the combining weights of zinc and oxygen, and hence the chemical force in the battery, that in the decomposing cell, and the electricity that links them together, are equivalent forces.

For another illustration of this conversion of force in definite quantities, notice the following curious results of careful experiments. An electric current of a given strength, when passed through water, will decompose it, giving a certain quantity of hydrogen. The gas thus obtained, when burned will produce heat by which the temperature of one pound of water may be raised *one degree;* but if the *same current* of electricity is changed to heat by passing through a fine wire, this heat applied to one pound of water will raise its temperature *one degree.*

4. *Motion is the medium of exchange among forces.*—We have seen that the *force* of gravity produces *motion*, and that the motion checked evolves the *force* of heat; the order is this: gravitation, *motion*, heat. By applying heat *force* the molecules of a bar of iron are put in more rapid *motion*, until it shines with a red

or even a white *light;* and in this case also we notice the order—heat, *motion*, light. Or again; the heat in combustion is but the manifestation of vibratory *motions* caused by chemical force. The order in which these occur is chemical force, *motion*, heat. And finally, to notice a more complex example, chemical force in the burning of coal through the *motion* of molecules is changed to heat. This heat, applied to water in the boiler of the steam-engine through the *motion* of the molecules of water, shows itself as expansive force of steam. This expansive force, through the *motion* of the piston and the wheels, is manifested as mechanical force to draw the train of cars.

Other examples might be given, but these must suffice. The truth they illustrate is this: All conversion of force from one form to another is accomplished through the medium of motion, either of molecules, or of masses of matter. It seems almost, then, as if we have a right to conclude that all forms of physical force are only so many different manifestations of motion. If we do not thus conclude, it is only because a theory so startling is not acceptable upon evidence which in ordinary matters would be thought sufficient.

But suppose the scientist should convince us that the "forces of nature" are *all* only different manifestations of motion; how constant, how complex, how variable, and yet how regular must these motions be! What keeps the molecules in continual, ever changing, and harmonious motions? We never think without amusement of the old philosophy which taught the existence of a flat earth resting upon the back of a huge elephant, himself standing upon turtles, but which left the turtles to support both themselves and their load; and yet what better is this modern science, if, after adroitly building itself upon molecular motions, it leaves us to suppose that the molecules move themselves! The "forces of nature" *may* be but different manifestations of molecular motions; it may be possible for science to prove this, but when this is done, science can go no further in this direction; for "who by searching can find out God!"

INDEX.

The numbers refer to the pages.

www.ingramcontent.com/pod-product-compliance
Lightning Source LLC
LaVergne TN
LVHW050527100826
845148LV00002B/471

* 9 7 8 1 4 2 5 5 1 9 9 8 8 *